U0291559

# 室内设计 风格详解 北欧

# MANUAL OF NORDIC INTERIOR DESIGN

徐士福　陈炬　张鋆　陈加强　主编

江苏凤凰科学技术出版社

# 第一章 Chapter 1

## 北欧风格的形成和发展 The Formation and Development of Nordic Interior Design Style

# 第二章 Chapter 2

## 北欧风格设计的元素和运用 The Application of Nordic Design Elements

## 第三章 Chapter 3

## 提升品位的北欧设计 Enhancing Northern European Interior Design

ARTICHOKE
ARTICHOKE
ARTICHOKE
GRAND CRU CLASSÉ DU MÉDO
CHÂTEAU
Bernard Magrez
ES CLÉS DE L'EXCELLENCE
ARTICHOKE
Bernard Magre
ARTICHOK

# I

第 1 章 Chapter 1

# 北欧风格的形成和发展

THE FORMATION AND CHARACTERISTICS OF NORDIC INTERIOR DESIGN

# 第一章北欧风格的形成和发展

Chapter 1
The Formation and Characteristics of
Nordic Style

## 一、起源 The Origin of Nordic Style

### 1. 风格定义
The Definition of Nordic Style

北欧设计风格是指起源于斯堪的纳维亚地区的设计风格，因此也被称为“斯堪的纳维亚风格”。北欧离欧洲中心国家较远，有着自己独特的气候和传统，在历史上北欧设计很少受到关注，直到 19 世纪末，北欧设计受到英国工艺美术运动和新艺术运动的影响，逐渐参与到各种设计运动中。20 世纪 20 年代，北欧设计师将北欧多面性的设计哲学融入现代主义设计的民主精神之中，卡雷·克林特（Kaare Klint）的家具探索、阿依努·阿尔托（Aino Aalto）的压制玻璃都奠定了北欧风格的发展方向。尽管北欧各国的设计都有着杰出的历史背景，但是在 1950 年的一次以“设计在斯堪的纳维亚”为主题的展览后才被作为一种风格而被广泛接受，这次展览从 1954 年到 1957 年在美国和加拿大巡回展出，使得“北欧风格”开始真正受到关注。

Nordic style indicates the design style originated from Scandinavian area, thus, also called Scandinavian style. As Northern Europe is far from the center of Europe, and it owns distinctive weather and traditions. Nordic style had received little attention until the end of 19th century when Arts and Crafts Movement had a great impact on it, and then it gradually became the participant in various kinds of design movement. In the 1920s, Scandinavian designers incorporate the design philosophy of Nordic versatility into the democratic spirit of modernism style design. Kaare Klint's exploration on furniture and Aino Aalto's pressed glass laid the design direction of the Nordic style.The design of the Nordic countries had a quite outstanding historical background, but it was not until an exhibition of "Design inScandinavia" in 1950 that was widely accepted as a style. It was a traveling exhibition from 1954 to 1957 in the United States and Canada, which had made Scandinavian style really being focused by the public.

北欧风格与装饰艺术风格、流线型风格等追求时髦和商业价值的形式主义不同，北欧风格简洁实用，体现对传统的尊重，对自然材料的欣赏，对形式和装饰的克制，以及力求在形式和功能上的统一。

Nordic style are different from Art Deco style or streamline style ,which are formalism of pursuing fashion and commercial value. The Nordic style is concise and practical, reflecting respect for tradition, appreciation of natural materials, restraint of form and decoration, and strive to achieve unity in form and function.

## 2. 区域界定
## The Location of Scandinavia

斯堪的纳维亚地区因地处欧洲北部的斯堪的纳维亚半岛而得名，通常指丹麦、 芬兰、冰岛、挪威和瑞典五个国家。而事实上只有挪威和瑞典两个国家真正坐落在斯堪的纳维亚半岛上，但由于这五个国家都位于欧洲的北部地区，且彼此邻近，因此，“斯堪的纳维亚地区”实际上指的就是北欧。正因为这种地域的临近，这些国家的历史背景彼此相互关联，语言相近（芬兰受俄语语系的影响较大，自成一派），因而这些国家几个世纪以来一直有很好的文化交融，并且建立了共同的意识形态，从而对于一种风格的形成奠定了良好的基础。

Scandinavia is a peninsula located in Northern Europe, generally refers to Denmark, Finland, Iceland, Norway and Sweden. In fact, only Norway and Sweden are the countries that situated in Scandinavian Peninsula. As these five countries are all located in Northern Europe and are adjacent to one another, Scandinavia became the alternative name of Northern Europe. As they are neighborhood in the aspect of geographic location, their historic background and languages are similar. Finland was influenced by Russia, thus different from the other four countries' languages. These five countries have well incorporation and shared ideologies, thus can build a sound foundation to forming the same style.

## 3. 人文风情
## Culture and Tradition

直到一万多年前北欧才摆脱冰川和大海的限制，成为欧洲最年轻的地区，事实上斯堪的纳维亚半岛本身就是在地球最后一个冰川期形成的，直至今日冰岛与斯堪的纳维亚山脉的面积依然被现代冰川占据了一部分。受北大西洋暖流、挪威暖流及西来气旋的影响，整个斯堪的纳维亚地区夏季温和而短促，冬季寒冷而漫长，属寒温带气候。由于斯堪的纳维亚是欧洲冰盖消退最晚的地区，林区树种相对贫乏，生态环境比较单一，但是森林面积广阔，是世界上重要的木材及木制品生产出口地区。在历史上北欧五国曾被当作一个整体，北欧诸国之间有着非常密切、休戚与共的姻亲关系和相似之处。这不仅仅是因为它们具有共同的自然背景，在更多情况下是由于它们有着共同的宗教信仰、类似的文化传统和相似的政治环境、经济模式等背景因素。

10,000 years ago Northern Europe was just extricated itself from glaciers and oceans. It is the youngest area in Europe. Scandinavia peninsula taken shape in the last period of Ice Age, up to now, still a part of the area in Iceland and Scandinavian Mountains is occupied by glaciers. Effected by the warm current of Norway and North Atlantic, and the cyclone from the west, the whole Scandinavia area has temperate and short summer, but cold and long winter. It belongs to cool temperature zone. In Scandinavia where ice sheets melt very late compared to the whole Europe, so the environment is poor for planting trees. The ecological condition is quite simple, and area of forest is board.

It is an important area of timber and woodworks exportation in the world. In history, the five countries in Northern Europe were regarded as a union. They had very close relations and shared jointly in-laws. It is not only because they have similar natural background, but the related culture, tradition and similar political background, economic pattern etc.

北欧是一个高福利、保障制度完善、注重休闲的地区。北欧人的自然、简单、幸福的慢生活态度是现代人所向往的，北欧的居民们回归平实、悠然惬意，被称为地球上最懂得生活的一群人。下午四五点钟，北欧人已经下班开着车到达湖边或山上，开始了他们的休闲度假。北欧国家并没有太多的高楼大厦，人们穿着朴素，开旧车，吃简单健康的食物，没有灯红酒绿的夜生活，也没有奢华的消费刺激人的神经。他们重视生活的品质胜于物质追求，重视与家人的团聚时光。但这并不意味着北欧人对工作的倦怠，为了提高工作效率，北欧人总是绞尽脑汁，思考如何改良技术，改进机器设备。

北欧人具有节俭及适度的消费观念，他们重视实际效用，但并不是降低自身的享受，而是将一些外在的浮华抛弃掉，正因为如此，北欧人比其他先进国家的人活得更舒适也更自由自在。北欧人讲究务实、不浮夸，不迷信名牌，只注重对象本身的价值，因此北欧家具以实用、耐用著称，售后的维修服务也较为完善。北欧人提倡购买经久耐用、兼具多种用途的家具及装饰品，甚至要考虑是否能够回收。

Northern Europe is an area enjoys high welfare, completed social security system and leisure environment. The residences living in this area are simple and casual. They are regarded are the group who enjoy life most on earth. When four or five clock in the evening, people will drive their cars to the lake or mountain, start their vocation. Northern European countries don't have many tall buildings. People wear plain clothes, driving old cars and eating healthy and natural food, without ritzy and glitzy nightlife and luxurious consumptions to stimulate their nerves. However, they are not bored with work. For enhancing the efficiency of work, Northern Europeans always rack their brains thinking about how to enhancing the technologies and machine equipment.

Northern Europeans hold moderate and thrifty consumption attitude. They pay attention to practical function, but not to minimize the enjoyment of life. It is discarding the superficial decorations. Therefore, Northern Europeans are more relaxed and freer than other people in developed countries. They stress on practical and unexaggerated essence, and are not obsessed with brands. They only emphasize on the inner value of the object itself, so Northern European furniture is renowned for it's practical and endurable function, and maintenance service is completed. Northern Europeans advocate purchasing endurable and high functional furniture and adornments and even consider whether they can be reclaimed.

# 二、特征 Characteristics

北欧风格以简洁著称于世，并对后来的“极简主义”、“简约主义”、“后现代”等风格产生影响。在20世纪全球风起云涌的“工业设计”浪潮中，北欧简洁风格被推崇到极致。北欧设计没有受到巴洛克、洛可可奢华风的影响，它始终保持了淳朴自然、简洁实用的设计风格。

Nordic style is known for its concision, and it had impact on the later "Minimalism", "Simplism" and "Post-Modernism". In 20th century, the world was in the wave of "Industrial Design", while the concise style of Nordic style was extremely highly praised. Scandinavian design was not subjectded to the luxurious wind of Baroque and Rococo. It always keeps the simpleand natural, concise and practical design style.

## 1. 简洁、通透的室内设计
## The Concise and Transparent Interior Design

北欧属于高纬度地区，冬季漫长且缺少阳光的照射，所以北欧人在室内空间设计上，最大限度地将阳光引进室内。正因为如此北欧的窗子规格较大，房间敞亮；色调往往以白色、米色、浅木色为主，而墙面的颜色，更是往往以纯净的白色为主；室内空间的格局设计没有过多的转折或拐角，这种空间结构简洁，大窗户的设计更有利于将阳光引入室内，白色的墙面则更有利于光线的反射，使房间显得更加宽敞明亮。

Northern Europe is a high latitude area. Winter is long and lack of sunlight, so Northern Europeans apply the design that willg reatly introduce the sunlight indoor. Thus, they have large size windows and the room is full of daylight. The colors are mainly based on white, beige and lightwood color, and the color of walls is usually in pure white color. There are no much corners or turns in the space structure, as it is more concise. The design of big windows is convenient to introduce the sunlight indoors, and white color of walls is propitious to the reflection of light, thus can make the room significantly transparent.

## 2. 室内空间以木材为主
Indoor Space Mainly Based On Wooden Materials

木材是北欧室内设计的灵魂，北欧的室内空间少不了木材的运用。北欧大部分地区处在北纬 61° ~ 68° 之间，是密集的针叶林分布区，瑞典更是世界上木材与木制品的主要生产与出口地，得天独厚的自然资源优势使北欧地区的室内设计有大量运用木材的条件。木材有利于室内保温，这非常适合在高纬度地区运用，北欧风格的室内装饰常使用未经过精细加工的原木，以体现室内整体的自然风格和质感。木材本身具有柔和的色彩、细密的天然纹理，在经过细加工组装变形后，成为制作各种家具的主材，例如枫木、橡木等。北欧地区的冬季漫长多雪，为了防止积雪过重压塌房顶，其建筑多以尖顶、坡顶为主，室内可以见到原木制的梁。

Wood is the spirit of Nordic design, because it is an indispensible material in indoor space. Most parts of northern Europe are between latitude 61 degrees to 68 degrees, they are dense zone of coniferous forest. Sweden is a productive country of timber and woodworks exportation. The unique natural resources provide Northern Europe with a large number of wooden materials. Wood is conducive to preserve heat, which is very suitable for high latitude area, the Nordic style indoor often use unrefined logs, in order to reflect the overall indoor natural style and texture.Wood owns soft color and natural texture. When it isrefined and assembled, such as maple and oak can be used as the main material for many kinds of furniture. The winter in Nordic region is long and snowy, in order to prevent the overweight snow on roof, the buildings are mainly with a spire, slope roof, and with original wooden beams indoor.

## 3. 风格自然、纯朴

Natural and Simple Exterior Design

北欧的室内设计装饰材料主要有木材、纺织品、玻璃、铁艺等，在实际运用中基本上都保留了这些材质的原始质感与肌理，例如直接堆砌在火炉边的原木、兽皮形状的绒毛地毯。这些都是室内自然风格的点缀，使室内空间充满人情味。北欧室内的天花、墙面、地面很少用纹样与图案来装饰，通常采用色块和线条区分。

The main decorating materials in Nordic interior design are wood, textiles,glass and ironwork etc. The original texture will be maintained while practically utilizing these materials, such as the randomly piled-up logs beside the fireplace, fur shape fluffy carpet. These ornaments with natural style fill the room with human kindness. The ceiling, walls and floor are rarely decorated with patterns. Instead, they generally divided by color blocks and lines.

北欧人因其得天独厚的自然环境，与自然相处十分亲密。斯堪的纳维亚地区森林茂密，湖泊星罗棋布，山脉高耸。同时，因为北欧高福利的生活状态，大部分人们不须过多为生活而劳累奔波，他们工作时间短，大部分时间用来进行户外运动和旅游。在进行室内设计时，北欧人把对自然的爱引入到室内空间中来，在家具中也体现了这种粗犷、自然的风格。而这些还与北欧人的生存环境与传统有着直接关系，远古的北欧民族，生活在严苛的自然环境中，他们的日常生活方式是狩猎、流浪、战斗，他们也会在大胆进取的首领的率领下远征他国，并从异国赢来财富与地位。北欧民族看似冷静，内心却火热，绝不优柔寡断，他们拥有纯粹、简洁的表达方式，这种性格特征自然也体现在室内空间设计上。

Nordic people are very close to nature as their distinctive natural environment. Scandinavian region enjoys dense forest, lakes and high mountains. As their high social welfare, most of them don't need to struggle for living. They spend most of their time to do outdoor exercise or traveling as they have short times for work. Their love for nature can be found in indoor, and furniture also manifest such natural and rough style. It has a direct connection to Nordic people's living environment and traditions. The ancient Nordic people were living in harsh natures. Their daily activities are hunting, wandering and fighting. They boldly marched for other kingdoms under enterprising leaders, and gained wealth and status from overseas. Nordic people are calm in appearance, but warm inside. They never show hesitation and they always keep true personality. Their ways of expression are direct and simple.Such personality and characteristics will also show in their interior design.

# 三、设计理念 Design Concept

北欧设计风格注重人与自然、社会，与环境的有机、科学的结合。这种风格集中体现了绿色设计、环保设计以及可持续发展的设计理念。它显示了对手工艺传统和天然材料的尊重与偏爱，在形式上更为柔和与有机，因而富有浓厚的人情味。

Nordic design lays stress on people's harmonious incorporation with nature, society, and environment. Such style manifests the concept of green design, environmental design and sustainable development. It shows respect and preference of handcrafts and natural materials. It is more soft and organic in form, and full of humanism.

## 1. 重视传统手工艺
## The Value of Traditional Crafts

斯堪的纳维亚地区诸国底蕴深厚的传统工艺是北欧现代设计的根源与基础。在工业化进程的早期，由于历史文化和地理因素，北欧并没有出现像欧洲大陆其他国家那样传统手工艺与机械化激烈对峙的局面。当现代设计风潮在全球盛行之际，北欧诸国对自己的手工艺传统更加小心地保护。在不断吸收现代设计思想的同时，努力使自己的设计不被世界潮流淹没，力求传统与现代的有机结合，在深厚的传统底蕴中积极探索，体现了简约、灵巧的现代精神，形成独特的设计风格。在家具设计方面北欧的经典案例数不胜数。这种设计文化的形成，除了源于北欧人固有的执着认真和敬业精神，更重要之处还是对时代潮流和设计趋势的准确把握与理解。从北欧设计中，我们强烈地感受到设计文脉与传统工艺的延续，以及和北欧人敢于不断创新和挑战新事物的创造精神。

十八世纪 60 年代挪威民间艺术手绘古董柜
Norway Folk Art Hand Painted 1860's Antique Cabinet

The sophisticated traditional crafts of Scandinavian region is the foundation and origin of Nordic design.In the early process of industrial development, as the historical and geographical factors, Northern Europe, unlike other countries in the Continent, did not experienced severe contradiction between traditional handcrafts and mechanism. When the world was in the modern design trend, Nordic countries were protecting their traditional handcrafts more carefully. While absorbing the idea of modern design, they tried to make their design not drowned in the world trend.They strived for the organic combination of tradition and modern, and keep exploring in the profound tradition. It reflects their simple and ingenious spirit, and the formation of unique design style.Nordic classical design casesare countless in Home Furnishing.The formation of this kind of design culture, in addition to the inherent dedication of Nordic people's serious, dedicated spirit, but also the important accurate grasp of the trend of the times and the understanding of design trend.From the Nordic design, we strongly feel the design context and the continuation of the traditional process, and their bold innovation and challenge of creating new things.

## 2. 以人为本、功能实用
People-Oriented, Functional and Practical

北欧的室内设计是源于生活的设计，无论是室内空间还是家具都体现以功能实用为主，同时美感创新和以人为本的设计理念。例如空间的简洁通透配合大窗的设计，就是最大化满足北欧人对阳光的需求。北欧设计既注重设计的实用功能又强调设计中的人文因素，同时避免了过于刻板的几何造型或者过分的装饰，恰当运用自然材料并突出自身的特点，开创了一种富于“人情味”的现代设计美学，受到人们普遍欢迎。

Scandinavian interior design is originated from dailylife. No matter interior space or furnishing, all presents its practical function, innovated aesthetic and people-oriented design concept. For example, the space of simple and transparent with large window , is to meet the Nordic people's needs of sunlight. They not only pays attention to the practical function design and emphasizes the human factors, but also avoid of too rigid geometric modeling or excessive decoration factors. Proper use of natural materials and highlight is the characteristics.It created abound of human touch, and is generally welcomed by the people.

## 3. 崇尚自然
Worship of Nature

在北欧的设计中，崇尚自然的观念是比较突出的，从建筑设计到室内空间设计以及家具的选择，北欧人都十分注重对本地自然材料的运用。以对浓厚的工艺技术尊重为前提，以人的设计为根本，辅以现代的人文功能主义的手法，借助高科技设计手段，北欧在工业化大生产模式下，靠温情浪漫、自然朴实的生活基调，立于不败之地。

In Nordic design, advocating natural concept is prominent. From architectural design to interior design and furniture selection, the Nordic people attach great importance to the use of local natural material. Based on the foundation of profound craftwork, and people-oriented design concept, they made their design simple but romantic. It withstood the impact of massive production pattern, succeeded to take a position in the design field.

Ⅱ

第 2 章 Chapter 2

# 北欧风格设计的元素和运用

THE APPLICATION OF NORDIC DESIGN ELEMENTS

# 第二章 北欧风格设计的元素和运用

Chapter 2
The Application of Nordic Design Elements

## 一、家具 Furnitures

### 1. 北欧家具的历史与发展
The History and Development of Nordic Furniture

北欧各国早期的家具设计较多受到德国的影响，但北欧自身拥有较深的传统工艺根基，在民间一直存在淳朴、实用的思想观念，成为北欧现代家具的根基。下面分别介绍丹麦、瑞典、芬兰和挪威北欧四国现代家具发展的情况。

The design of Nordic furniture was influenced more by Germany, but Scandinavia owns deep traditional craftwork background itself. There are always practical and natural ideas exist in Scandinavia, thus, they have become the bases of Scandinavian furniture. The development of the four countries in Scandinavia, including Denmark, Sweden, Finland and Norway, will be illustrated in the following text.

#### (1) 丹麦 Denmark

早在四百多年前，丹麦就有了同行业的组织“基尔特”。这些组织经常将成员们聚集到一起，相互交流制作技艺，这个良好传统一直保持到现在，这为丹麦家具的传承和发展起到了积极的促进作用。丹麦在 1907 年成立了工艺设计协会，它对手工艺技术的提高、产品实用化的促进做出了较大的贡献。但真正深入研究并奠定了丹麦家具设计基础的是设计大师凯尔·克林特，大部分丹麦家具在克林特之前是照搬欧洲家具的样式，这无论从实用还是从民族传统艺术传承的角度来看都是不可取的。作为一个建筑师的克林特，通过把实用技术和艺术相结合，将桌子和椅子的高度以及抽屉尺寸标准化，提倡使用自然材料，色彩为单纯的黑色、白色、灰色或棕色，为丹麦家具的形式与内涵翻开了新的一页。随后在克林特的影响下和感召下，涌现出了极具个性、自成一家的设计大师，如：芬·居里和最能代表丹麦风格的汉斯·韦格纳以及阿诺·雅各布森。

In 400 years ago, Denmark had formed many design groups. They were named "Guild". The groups often gather together to exchange their ideas of design craftwork. The tradition has been remained until today, and it has greatly promoted the inheritance and development of Danish furniture. In 1907, Denmark established an arts and crafts design association. It contributed great enhancement to handcraft and product practicality. However, in real in-depth study, the one who laid the foundation of Danish furniture design is Kaare Klint. Most of the furniture was copying the style of European furniture before Kaare Klint's design. It was not desirable neither from practical nor ethnic aspect. As an Architect, Klint made a combination of practical technology and art, and standardize the height of table, chair and the size of the drawer. He promoted natural materials, and colors of pure black, white, gray or brown. Since then, the content of Danish furniture opened a new page. Under the influence and inspiration of Clint, emerged some unique master designers, such as Finn Juhl, and Hans Wegner, and the most qualified representative of Denmark style, Arhe Jacobsen.

## (2) 瑞典 Sweden

瑞典家具同丹麦家具一样在世界上有着很高的声誉，瑞典早在1845年就成立了工业设计协会，北欧的产业革命在欧洲是相对落后的，但瑞典的产业革命在北欧起着引领的作用。1910年以后，瑞典开始重视对生活日用品的设计，提出“选择更好的用具，可以改变生活的式样”等口号，以此来强调设计的重要性，号召设计师、艺术家深入到工厂一线，设计出更完美的日用产品。1930年瑞典在斯德哥尔摩举办了一次有关住宅的展览，参展的作品无不兼具实用功能与造型之美，在全世界引起了轰动。这也为北欧的家具设计行业创造了新的机遇，瑞典在全国大力推行家具设计运动，并且取得了显著的成就，在1937年的巴黎展览会和1939年的纽约展览会上都曾获得极高的评价，自此，瑞典家具在国际上占有了一席之地。

第二次世界大战后，一位名叫欧凯尔·布隆的医生发表了一篇《座椅与工作效率》的研究报告，引起了国际上的重视，这项研究成果奠定了人体工程学在家具上研究与运用的基础。与丹麦家具相比较，瑞典的家具以实用性为主，致力于兼具工艺性与市场需求的大众家具的研究与开发。瑞典家具商也曾受到丹麦的一些影响而采用温热带木材如柚木、紫檀等。

Swedish furniture enjoys the same high reputation as that of Denmark. Sweden had founded an industrial design association in 1845. The industrial revolution in Northern Europe is relatively backward in Europe, but the Swedish industrial revolution in Northern Europe plays a leading role. After 1910, Sweden began to attach importance to the design of daily necessities of life. It advocated the slogan of "choose the better equipment, change the style of life." so as to emphasize the importance of design. It called on designers and artists to go deep into the factory to design more perfect daily necessities. In 1930, Sweden held a house exhibition in Stockholm. The works were both practical and beautiful. It caused a sensation in the whole world, and also created new opportunities for the Scandinavian furniture design industry. Sweden had greatly promoted the movement of furniture design and had made remarkable achievements. The Swedish furniture design in the Paris exhibition in 1937 and New York exhibition in1939 had been greatly evaluated. Since then, the Swedish furniture gained an international status in the world.

After the Second World War, a doctor named Okel Blom published a research which report a study of the seat and work efficiency, and it attracted international attention. The research has laid the foundation for the research of application of furniture on the effect of human body. Compared with the furniture of Denmark, Swedish furniture is more practical, and is devoted to the research and development of the technology and market demand. The Swedish furniture business was also affected by Denmark, so they applied wood from warm and tropical zone, such as teak and rosewood.

## (3) 芬兰 Finland

芬兰起初长期被瑞典占据，后又被沙俄侵占了百年，直到1917年才正式独立。芬兰作为一个年轻的国家，在各方面都显得生机勃勃。早在1875年芬兰就有了自己的工艺设计协会，但真正发扬芬兰的艺术风格，将湖光山色的独特元素融入设计作品之中，却是在1910年前后。在芬兰著名的建筑设计师沙利宁等的领导下，加之北欧人在造型上特有的天赋，芬兰的家具设计行业很快就形成了一定的规模，后来芬兰的著名建筑设计师阿尔瓦·阿尔托为芬兰家具带来了无限生机。芬兰家具的最大特色是在设计上强调个性，同时又利用国内现有材料进行大量生产。

In the beginning, Finland had been occupied by Sweden for a long time, and then invaded by Tsarist Russia. Until 1917, it was formally independent. As a new country, Finland is full of vitality. As early as 1875, Finland had its own art crafts design association, but not until around 1910 had the real development of Finland art style which integrates unique elements of landscape of lakes and mountains into the design works. In Finland's famous architectural designer Saarinen Eero and other leaders, coupled with the Nordic people in the form of a unique talent, Finland's furniture design industry will soon form a certain scale, and later Finland's famous architectural designer Alvar Aalto brought Finland with the unlimited vitality of furniture. The biggest feature of Finland's furniture is to emphasis on personality, while usingof domestic materials to make mass production.

## (4) 挪威 Norway

挪威的家具朴素而充满乡村气息，这与挪威的风土人情十分吻合。挪威的设计协会成立于1938年，因为创立的时间最晚，所以挪威人可以借鉴北欧其他国家的经验。他们借鉴丹麦、瑞典等国家在家具设计中的优秀成果，融合自己的风俗与民情并加以发展创新，形成了挪威自己的独特风格。挪威的家具设计，无论在胶合板材还是金属等材料上都能匠心独运，创造出风格独特的产品。现在挪威家具的设计大致可分两类，一类是以出口为目的，在技术与材料上都极具特色。另一类是自然、朴素具有本土气息的供国内日常家庭使用的家具。在挪威有许多著名的家具设计师，如以独特优雅家具而著称的托尔比昂·阿伏达；以浓郁的挪威民族风格为特色的比昂·伊安克等。

Norwegian furniture is simple and full of local flavor.It is consistent with the local customs and practices. Norway's Design Institute was established in 1938. Since the establishment is the latest in Scandinavia, the Norwegian could learn from the experience of other countries. They borrowed the great achievements from Danish and Swedish furniture, and incorporated with their local feature for an innovation. They formed their unique characteristic. The furniture designs in Norway are excellent and creative no matter in the plywood or metal materials. It has created special productions. Now Norway furniture design can be divided into two categories, one is for the purpose of exportation. It is highly characterized in the technology and materials. Another one is natural and simple domestic family furniture with the local flavor. In Norway there are many famous furniture designers, such as Torbio Avoda is known for his unique elegant furniture, and Bion Yiankewith rich ethnic characteristics.

# 2. 名师名作
Masterpieces

## (1) 凯尔·克林特 KaareKlint

凯尔·克林特是丹麦现代设计派的开山鼻祖，被誉为丹麦现代家具之父。1924 年，克林特组建了哥本哈根皇家艺术学院的家具设计系，并担任该系的教授及系主任。自此，丹麦设计学派形成并得到了飞速发展。早在 1916 年，克林特就主张家具设计的简洁化和实用性，现在的丹麦家具也是以这两种设计思想而闻名于世。后来为了使家具更具有舒适性和实用性，他曾对人体解剖学和人体肌肉的活动进行了详细的研究，他还是最先使用未上清漆的木料和自然织物的人之一。

Kaare Klint is the founder of Danish design. He has been viewed as the father of Danish modern furniture. In 1924, Klint organized a furniture design department of Royal College of Art in Copenhagen, and served as a professor and head of the Department.Since then, the Danish design school has established and developed rapidly. Early in 1916, Klint advocated the simplicity and practicality of furniture design, and now the Danish furniture is also known as these two design ideas. In order to make furniture more comfortable and practical, he did a detailed study of human anatomy and human muscle movement. He is the first person to use the wood of no varnish and natural fabrics.

克林特并不随波逐流，他没有淹没在“国际主义”的冷漠之中，反而走了一条全力挖掘传统的道路。在克林特的“传统”中对英国乡村家具、中国明清家具颇为重视，克林特早期就意识到，家具设计的主要问题就是运用前人作品中的精华进行现代设计。“传统”不论古今，不论国别，只要认为合适，就可以拿来研究和改进，最后创造出自己的作品。他从各国文化出发研究了众多家具，仔细分析了它们的比例和尺寸，博采众家之长以设计出实用、美观的现代家具。他的这些设计方法激发了后人对“功能主义”家具的兴趣，在 20 世纪 30 年代被其他国家的设计师广泛采用。

Klint did not follow the crowd, and he did not drown in the "internationalism" styles. On the contrary,he did a full excavation on traditional road.In Klint's road of tradition, he paid attention to the furniture of United Kingdom and the Ming and Qing Dynasties in China. Klint had realized early that the main problem of the furniture design is the use of the essence of previous works in modern designing. Regardless of ancient, modern or countries, as long as it is suitable, it can be used for study and improvement, and finally to create their own works.He did researches on a lot of furniture from the aspect of various cultures in different countries. He carefully analyzed their size and proportion and gathered their unique excellence to design elegant and practical furniture. His design method had inspired the later generations of the"functionalism". It was widely used by designers in other countries in 1930s.

克林特引导学生系统地研究室内设计和家具的理论知识、生理学及人体工程学等，为丹麦现代设计的崛起打下良好的基础，并直接或间接地培养了一大批丹麦现代家具设计的大师级人物，如：奥尔·温谢尔、布吉·莫根森、汉斯·韦格纳等。

Klint guided the students to systematically analyze the theoretical knowledge of interior and furniture design, physiology andergonomics etc,for laying a sound foundation on Danish modern design. It had directly or indirectlycultivated a large number of Danish modern furniture master designers, such as, Ole Wanscher, Borge Mogensen and Hans J. Wegner etc.

## Faaborg 椅

发表年份：1914

材料：桃花心木、藤

这款椅子是专门为法堡博物馆设计的，并在1914年－1923年间生产了大量的变形设计。

## Faaborg Chair

The year of coming out: 1914

Materials:mahogany, rattan

The chair was designed by Kaare Klint for the FAABORG museum. A large number of deformed designs produced during 1914 to 1923.

---

## 红椅（巴塞罗那椅）

发表年份：1927

材料：桃花心木、皮

红椅因在1929年巴塞罗那博览会上获奖，所以又称巴塞罗那椅。这款椅子的设计很明显受到了英国著名家具大师齐宾泰尔的影响。

## Red Chair or Barcelona chair

The year of coming out: 1927

Materials: mahogany, leather

The chair is also named Barcelona chair as it had rewarded in 1929 Barcelona Expo.This chair is obviously influenced by the design of the famous British furniture master Thomas Chippendale.

---

## 折叠式躺椅

发表年份：1933

材料：柚木、藤条、帆布

椅子搁腿的部分可以折叠至座位下，既可以躺着也可以坐着，而且座椅本身也可折叠，方便储藏。

## Deck Chair

The year of coming out: 1933

Materials:teak, rattan, canvas

Chair legs can be folded up under the seat, and can be either lying or sitting. The seat can be folded to put into facilitate storage.

### 螺旋凳

发表年份：1927

材料：岑木、帆布

螺旋凳取名于其独特的腿部曲线造型，当这张凳子折叠起来后，腿部形成圆柱状。螺旋凳的造型成功在于把简单的凳腿通过丰富的曲面而表现得与众不同。

### Propeller Stool

The year of coming out: 1927

Materials: ash wood, canvas

The chair is named after its unique leg curve shape. When the stool is folded, it will form cylindrical shape legs. The success of this modeling lies in the simple stool legs through the rich surface and the extraordinary performance.

---

### 安乐椅

发表年份：1932

材料：桃花心木、皮、藤

安乐椅的扶手与腿部一体结合，前腿直，后腿弯曲，既强调整体舒适感又体现个性设计。

### Easy Chair

The year of coming out: 1932

Materials: mahogany, leather, rattan

The arms and chair legs are integrated with the front legs straight and bent in the back. It not only emphasized the overallcomfortable feeling but also the character of design.

---

### 旅行椅

发表年份：1933

材料：岑木，帆布

旅行椅的原型是英国军队在印度殖民地使用的椅子，造型极为简单，方便携带。瑞典设计师 Arne Norrel 正是由于受到此款椅子的影响，后来才设计了一款相似的椅子。

### Safari Chair

The year of coming out: 1933

Materials: ash wood, canvas

The prototype of the Safari chair was used in the British army in India.It is very simple and easy to carry. Swedish designer Arne Norrel designed a similar chair after the influence of this chair.

## (2) 阿尔瓦·阿尔托 Alvar Aalto

阿尔瓦·阿尔托是芬兰现代建筑师，人情化建筑理论的倡导者，同时也是一位设计大师及艺术家。

Alvar Aalto is a Finnish modern architect. He advocates humanism architectural theory. He is also a master designer and an artist.

阿尔瓦·阿尔托与他的夫人——设计师阿诺·玛赛奥(Aino Marsio)共同进行了长达5年的木材弯曲实验，这项研究促使了阿尔托20世纪30年代革命性设计的产生。

Alvar Aalto and his wife had carried a 5 years experiment on bending wood. The research promoted the emergence of Aalto's innovative design in 1930s.

1929年，阿尔瓦·阿尔托运用功能主义建筑思想与他人合作设计了为纪念土尔库建城700周年而举办的展览会的建筑。他抛弃了传统风格的一切装饰，使现代主义建筑首次在芬兰呈现，推动了芬兰现代建筑的发展。1931-1932年，阿尔托设计了芬兰帕伊米奥结核病疗养院，一起在那里亮相的还有他最初设计的现代化家具，这是阿尔托的家具设计走向世界的重要突破。

In 1929, Alvar Aalto adopted functionalism architecture ideas and associated with others designed a 700 anniversary exhibition building for commemorating the building of Turku city.He abandoned all the decorations in traditional style and presented the modernism architecture for the first time in Finland. It had pushed the development of Finland's modern architecture. In 1931 to 1932,Alto designed the Sanatorium Paimio in Finland, and other modern furniture indoors. This is a significant breakthrough for Alto's furniture to gain an international popularity.

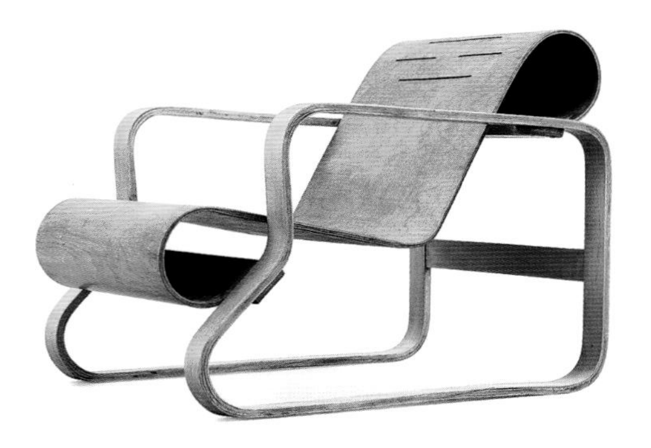

### 帕米奥椅

发表年份：1932

材料：层压胶合板

“帕米奥椅”是为“帕米奥疗养院”设计的。这件简洁、轻便又充满雕塑美的家具是由一整块薄板和塑造好的夹板共同拼合而成。它特征最明显的圆弧形转折并非出于装饰，而完全是结构和使用功能的需要，椅背上部的三条开口是为使用者提供的通气口。

### Paimio

The year of coming out: 1932

Material:Laminated plywood

Paimio was designed for PaimioSanatorium. This concise and light chair is full of sculpture shape, as it is formed by a slight board and shapedplywood. The most obvious feature of this chair is its circular arc line. It is designed for the demand of function and well structure. The three openings on the upper part of the seat back are used to provide ventilation for user.

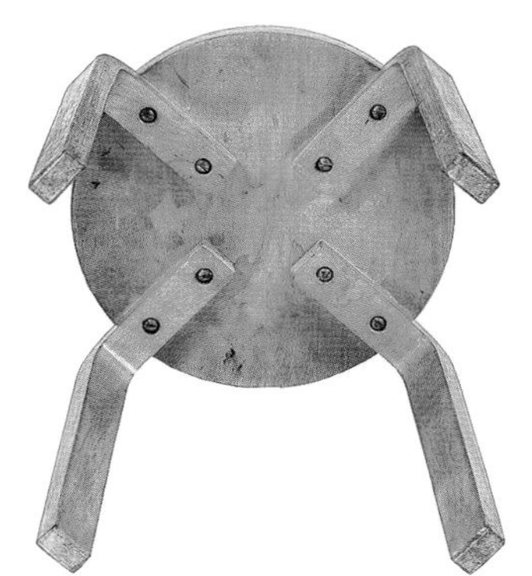

## 圆凳

发表年份：1932

材料：层压胶合板

阿尔托为维堡图书馆设计的一种叠落式圆凳，其最惊人的特点就是后来被称作“阿尔托凳腿”，面板与承足的连接，圆凳只有四个极为简单的构件，而叠落所形成的三重螺旋轨迹本身又构成了一件有趣的雕塑艺术品。圆凳设计的尺度、比例均可依具体场合的使用需要进行调整，同时亦可加上或高或低的靠背变成普通椅或酒吧椅。

## Viipuri Stool

The year of coming out: 1932

This chair was designed for Viipurili library.The stunning feature of this stool is called " Aalto leg ", as its special junction of board and the legs. The round stool has only four simple components, and the formation of the three-spiral trajectory is an interesting sculpture art. The size and proportion of this stool can be adjusted to different occasions. The legs can be lengthened and a chair back can added to displayed in the bar.

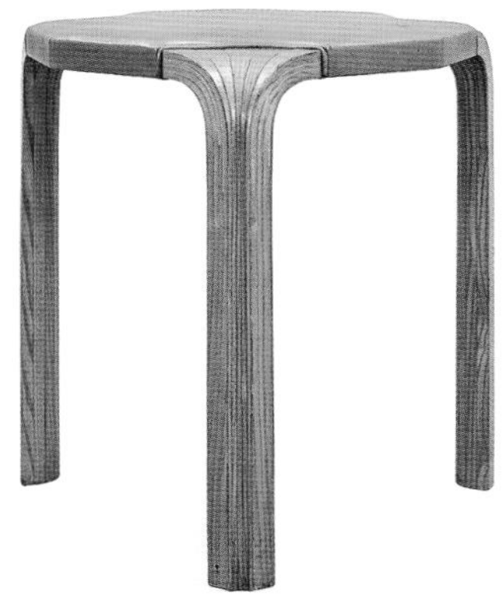

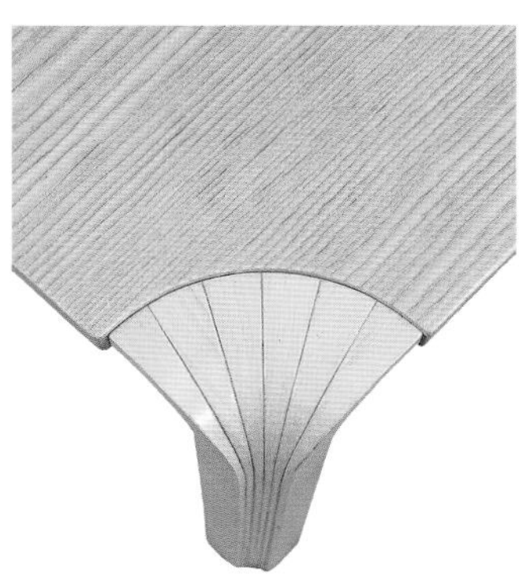

## 扇足凳

发表年份：1954

材料：桦木

扇足凳吸取了扇贝形态的元素，在凳腿与坐面的结合处做了巧妙的设计，装饰简洁而效果突出，凳腿线条的处理使整个圆凳更加挺拔，更加自然。

## Fan-leg Stool

The year of coming out: 1954

Material: birch

Scallop inspired the design of this round stool. The junctionof legs and surface is an ingenious design presents a simple decoration and outstanding impression. The processing lines make the stool and legs more upright, and natural.

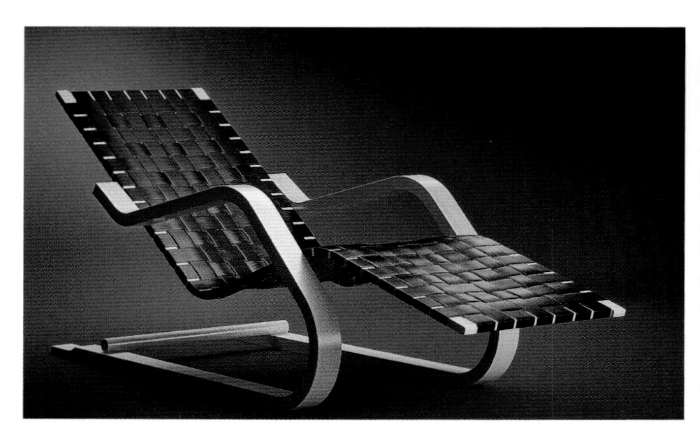

**层压胶合板悬挑椅**

发表年份：1931－1932

材料：层压胶合板，皮带

1938 年问世的层压胶合板悬挑椅是阿尔托在家具设计领域取得的又一伟大成就。马特·斯坦于 1926 年设计出第一把悬挑椅，从那时起，人们就误认为钢材是唯一能用于这种结构的材料。然而，阿尔瓦·阿尔托却在反复实验后确信层压胶合板也具备这样的性能，并成功地设计出了这件世界家具史上的第一把层压胶合板悬挑椅。

## Armchair 31

The year of coming out: 1931-1932

Material: Laminatedplywood, leather belt

In 1938, the invention of this advent of laminated plywood hanging chair is a great achievementin the field of furniture design for Aalto. Stam Mart in 1926 design a hanging chair, from then on, people mistakenly believe that steel is the only material that can be used for chair structure. However, after several experiments, Alvar Aalto convinced that the laminated plywood also can achieve such performance, and successfully designed the world's first laminated plywood cantilever chair in the furniture history.

---

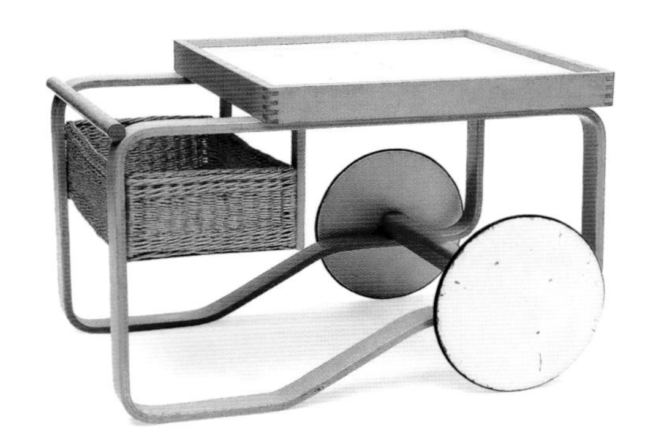

**茶盘车**

发表年份：1936－1937

材料：层压胶合板

阿尔瓦·阿尔托为阿太克公司设计的手推车式茶几，它采用光滑弯曲的胶合板制造，没有使用任何国外的标准组件，例如像螺钉和可拆装的连接件。整个作品很简单，即使是扶手也仅用一根硬木圆杆便制作而成。

## Tea Cart

The year of coming out：1936-1937

Material:Laminated plywood

This Tea Cart was designed for Artek Company. It is made of smooth and curved plywood, without using any foreign standard components such as screws and removable connectors.The entireproject is very simple, even its handrail is made ofa single hardwood round bar.

## (3) 雅各·布森 Arne Jacobsen

雅各·布森是20世纪最具影响力的北欧建筑师和工业设计大师，被誉为丹麦国宝级设计大师，北欧的现代主义之父，是“丹麦功能主义”的倡导人，他与勒·柯布西耶、密斯·凡德罗、阿斯普朗德等欧洲设计师共同主张简约风格设计。他不仅是20世纪最伟大的建筑师之一，同时在家具、灯饰、纺织品等应用艺术上皆有独到的见解与成就。

雅各·布森将自身对建筑的独特见解延伸至家饰品，他亲自为他的建筑内部设计家具装饰品，创造出诸如蚂蚁椅、蛋椅、天鹅椅等旷世之作，以及雅宅、工厂、展示间、纺织品、时钟、壁灯、门把手等多元创作作品。雅各·布森的设计将自然流畅的雕刻式造型与北欧设计的传统特质加以结合，使其作品兼具造型美观与结构完整的特色。他的作品至今依然保持吸引力，且深得人们的喜爱。

Arne Jacobsen was the most influential Nordic architects and industrial design master in 20th Century. He was Known as the Danish National Design Master, the father of the Nordic modernism, and is the advocator of "Danish functionalism". Le Corbusier,Mies van der Rohe, GunnarAsplund, Jacobsen and other designers in Scandinavia all advocated a concise style design.He is not only one of the greatest architects of 20th century, but also have unique insights and achievements in the furniture, lighting, textiles and other applied arts.

Arne Jacobsenextended his unique insight to other decorating, and designedfurniture decorations for the architectural interiors, such as Ant Chair, Egg Chair, Swan Chair and , and multiple creative works such as houses, factories, showroom, fabric, wall clock, door handle multiple works.Arne Jacobsen's design combined the natural flow of sculpture and traditional characteristics of Scandinavia, so that his works are beautiful and enjoy a characteristic of structural integrity. Until now, his works still remain attractive and are deeply loved by people.

### 蛋椅

发表年份：1958

材料：聚亚安酯、铝、皮

这把椅子的椅身、后背和扶手是由一整块聚亚安脂制成，整个造型取自蛋的形状，形体线条简洁流畅。使用者坐进去后整个身体被包在椅子里，有种被拥抱的温暖感，非常舒适。

### Egg Chair

The year of coming out: 1958

Material:polyurethane,aluminum, leather

The chair is made of a whole piece of polyurethane, the shape is preformed as an egg shape. The lines are concise and smooth. One would feel comfortable and wrapped in this chair when sits in. It creates a feeling of surrounding by warm arms, that make people relax.

## 天鹅椅

发表年份：1958

材料：聚亚安酯、铝、织物

天鹅椅的造型形象生动，宛如一只正在湖面上悠闲戏水的天鹅，故得名天鹅椅。天鹅椅后来也被投入到天鹅长沙发的造型之中。

## Swan Chair

The year of coming out: 1958

Materials: polyurethane, aluminum, fabric

This Swan Chair makes people associate with swans. Swan chair is also put into the shape of the Swan long sofa.The vivid shape of Swan chair like a bird on the surface of the swan, therefone named Swan chair.

## 蚂蚁椅

发表年份：1951

材料：榉树木、不锈钢

这把椅子是因其外形酷似蚂蚁而得名，虽然现在为了更稳定做成了四条腿，但最开始时是设计成三条腿的。

## Ant Chair

The year of coming out: 1951

Materials: Schneider zelkova wood,stainless steel

The shape of this chair resembles ants, that is why it was named Ant Chair. At the beginning of the design it had three chair legs, but concern of the stability, it changed to four legs.

## 大奖赛层压椅

发表年份：1957

材料：层压模制胶合板

这把椅子的椅身和腿都是由层压木板制成，其后背的造型比蚂蚁椅棱角更加分明，椅身曲线让使用者落坐时更舒适。

## Grand Prix Chair with Wood Legs

The year of coming out: 1957

Material: Laminated molded plywood

The chair and legs are made of Laminated board. The shape of the chair back is more distinct than Ant Chair. The curve line makes people to feel more comfortable when sits in.

---

## 壶椅

发表年份：1959

材料：聚亚安酯、钢管、织物

这款椅子因造型和有趣的名字而闻名。

## Pot Chair

The year of coming out: 1959

Materials: polyurethane, steel tube, fabric

This chair is famous for its interesting name and shape.

## 水滴椅

发表年份：1958

材料：聚亚安酯、钢管、皮

水滴椅的创意源于自然界水滴的形态，被设计为酒吧椅，它的椅身、后背同样是由一整块聚亚安脂制成，造型简洁，自然流畅。

## Drop Chair

The year of coming out: 1958

Materials: polyurethane, steel tube, leather

The shape of this chair inspired by the shape of water drops. It was designed to fit in the bar. The chair and the chair back are made of a whole piece of polyurethane. The shape is concise and the lines are natural and smooth.

---

## 七椅

发表年份：1955

材料：层压胶合板、钢

七椅是雅各布森的代表作之一，后来产生了很多变形设计，以满足不同的需求。这款椅子从设计到生产整整带动了一个行业，深受大众的欢迎。

## Seven Chair

The year of coming out: 1955

Materials: Laminated plywood, steel

This chair is one of the masterpieces of Jacobsen. It has many deformation design to meet different requirements. The design and production of this chair stimulated a whole industry. It had gained wide popularity.

## (4) 布吉·莫根森 Borge Mogensen

提到 20 世纪丹麦家具设计的现代化与大众化时，必然会提到丹麦消费者合作社 FDB 所属的家具设计部门，而它的主创者正是布吉·莫根森。作为凯尔·克林特的优秀学生，莫根森具有超出常人的创造能力，他设计了诸多造型简练优美、价格亲民的家具，例如著名的 J39 及 Spoke-back Sofa。

Refer to the modernization and popularization of furniture design in 20th century, the furniture design apartment in Danish consumers cooperative FDB will not be neglected. And Borge Mogensen is the founder. As one of the excellent students of Kaare Klint, Mogensen had distinctive creativity. He created a lot of simple and elegant furniture, and the price is affordable, such as the J39 and Spoke-back Sofa.

在 1950 年秋季以狩猎小屋为主题的丹麦家具匠师公会的展览中，布吉·莫根森以线条优美、大气的猎椅颇受好评。1958 年在游访了西班牙之后，他设计出了具有北非伊斯兰风格的作品——西班牙椅，成为莫根森的另一件代表作。莫根森的风格注重功能主义，并且体现传统与现代的结合。他主张好的家具不应该是少数人的玩物，而应该让人人都能拥有，借助工业化量产实现价格平民化，他的这一主张在丹麦乃至全世界都仍然发挥着影响。

In the fall of 1950, the exhibition of Danish Copenhagen Cabinet-makers Guild had the theme of hunting house. Borge Mogensen's elegant and splendid Hunting Chair was highly praised. After his traveling in Span in 1958, he designed a Spanish Chair with full of North African Islamic style. It had became his another representative work. Mogensen paid attention to functionalism, and the presentation oftraditional and modern combination. He advocated that furniture should not be the play things of few people, but should belong to everyone. With the help of efficient industrialization, he achieved the common price of furniture. His concept is still influencing in Denmark and the whole world.

**猎椅**

发表年份：1950

材料：橡木、皮

这款椅子的主要特点是采用皮带系紧结构来绷紧坐面和靠背。

### Hunting Chair

The year of coming out: 1950

Materials: oak, leather

The characteristic of this chair is using tightened leather as a belt to create seat and chair back.

### 西班牙椅

发表年份：1958

材料：橡木、皮

这张隐含着浓郁北非伊斯兰风格的椅子，造型厚重、粗犷。

### Spanish Chair

The year of coming out: 1958

Materials: oak, leather

This heavy and crude chair embedded a strong North Africa Islamic style.

---

### 布吉·莫根森两人位沙发

发表年份：1962

材料：实木、海绵、皮

布吉．莫根森两人位沙发是设计师莫根森在 1962 年为自己的家设计的，双人位沙发拥有简单的线条，饱满的座垫和精美的做工，扶手和靠背的高度确保了客户最大的舒适度。

### Borge Mogensen Two Seater Sofa

The year of coming out: 1962

Materials: solid wood, sponge, leather

Mogensen designed this sofa for his own house in 1962. It has simple lines, fluffy cushions and delicate handcraft. The height armrest and sofa back are designed to provide the most comfortable feeling when sits in.

## (5) 汉斯·韦格纳 Hans J.Wegner

丹麦设计师汉斯·韦格纳是20世纪最伟大的家具设计师之一，他一生作品超过500种，有的作品现在来看也是前卫而时尚的，是全球公认的最具创造力且多产的家具设计师。丹麦当今木制家具的声誉应首先归功于汉斯·韦格纳的设计，另外他还设计了很多灯具、墙纸、银器等。

Danish designer Hans J.Wegner was one of the greatest furniture designers in 20th Century. His works are numerous, more than 500 kinds, and some of his works are still vant-garde and fashionable today. He was the world's most creative and prolific furniture designer. Nowadays, the reputation of Danish wooden furniture should first attribute to Hans Wagner's design. In addition, he also designed many lamps and lanterns, wallpaper, silverware and so on.

汉斯·韦格纳早年潜心研究中国家具，1945年设计的系列中国椅就吸取了中国明代圈椅的精华。汉斯·韦格纳的设计尊重传统，崇尚自然，包含了一种“人情味”的现代美学。他设计的椅子结构科学，充分发挥材料特性，体现造型优美、细节完善、舒适简朴的风格特色，被人们称为椅子大师。汉斯·韦格纳非常看重手工艺，追求完美的手工艺雕琢打磨的细腻感。他的作品很少有生硬的棱角，转角处一般都处理成圆滑的曲线，给人以亲近之感。

Hans Wegner did a deep research on Chinese furniture in early years. The China Chair designed in 1945 was absorbed the essence of the chairs in Chinese Ming-Dynasty. The designs of Hans Wegner respect tradition and nature, and embraced an esthetic of humanism. The design and structure of his chairs is very scientific, and thoroughly illustrated the specialty of materials. The shapes are elegant and details are delicate. All present the comfortable and simple feature. He was called chair master. Hans Wegner pays attention to handcraft. He pursuit perfection and delicate sense of exquisite craftsmanship.His design rarely has blunt edges, and the corners are generally treated as smooth curve, for creating a sense of intimacy.

**中国椅**

发表年份：1944

材料：樱桃木或桃花木、皮革

这是韦格纳中国系列的第一把椅子，是在吸取中国传统圈椅元素的基础上，融合北欧的简洁格调设计而成。

**China Chair**

The year of coming out: 1944

Materials: cherry wood, mahogany, leather

This is the first chair of Wagner's Chinese series. It based on the Chinese chair's traditional elements, and a simple integration of the Nordic design and style.

## Y 型椅

发表年份：1950

材料：橡木、纸绳

这把椅子也是以中国传统圈椅为原型，巧妙地融合北欧的有机曲线，把弧度美到了极致。用"Y"字形的椅背，取代厚重的传统中式椅背，让中国沉重的太师椅，变得轻巧而灵活。

## Y chair

The year of coming out: 1950

Materials: oak, paper string

This chair is also based on the traditional Chinese armchair, skillfully combined Nordic organic curves. It presents the beauty of extreme bending. The " Y " shape chair back replaced the traditional Chinese chair back to turn the old fashioned armchair to light and flexible chair.

---

## 椅

发表年份：1950

材料：柚木、皮革

这把椅子是韦格纳的代表作，被称为"世界上最漂亮的椅子"。这把椅子最初是为有腰疾的人设计的，坐上去十分舒服，它拥有流畅优美的线条，高雅的造型和精致的细部处理。

## The Chair

The year of coming out: 1950

This chair is Wegner's masterpiece, known as the world's most beautiful chair.This chair was originally designed for people with waist disease. It is very comfortable when sits in. It has smooth and elegant lines ,exquisite detail processing, elegant and simple shape.

**孔雀椅**

发表年份：1947

材料：岑木、柚木、纸绳

汉斯·韦格纳最独特的椅子之一。这把椅子的外形酷似孔雀的形态，表现了他早期对仿生设计的探索，也表现了他丰富的创造力和以自然为本的设计理念。

## Peacock Chair

The year of coming out: 1947

Materials: ash wood, teakwood, paper string

This is the most special chair of Hans Wegner's design. The shape of this chair looks like peacock, which shows his early exploration of bionic design and also shows his rich creativity and natural design concept.

---

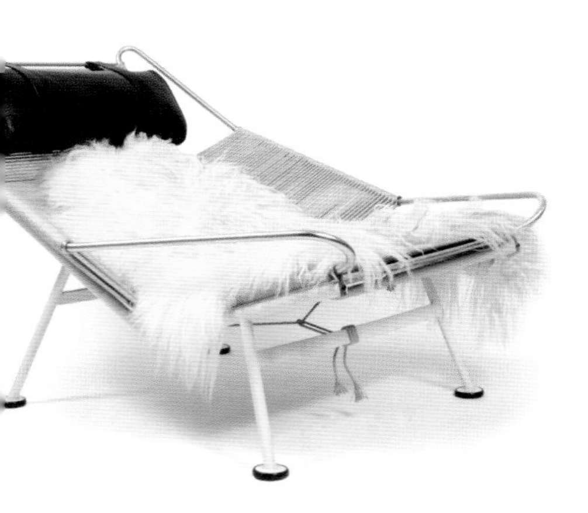

**旗绳椅**

发表年份：1950

材料：镀铬钢管、纸绳

这是一把有设计史料价值的躺椅。汉斯·韦格纳向现代主义大师勒·柯布西耶、密斯·凡德罗和马塞尔·布劳耶学习，用现代主义的手法来设计，使用现代的镀铬钢管做框架，旗索绳编制座位，羊皮来增加柔和舒适度。

## Flag Halyard Chair

The year of coming out: 1950

Materials: chromiumplated steel pipe, paper string.

The design of this chair has historical value. Hans Wagner used modernism approach on the design. It learned from modernism master Le Corbusier, Mies van der Rohe and Marcel Breuer.Using modern chrome plated steel tube as the frame, and knitted rope arrangement of seats, sheep skin to increase soft comfort.

## 挂衣椅

发表年份：1953

材料：橡木

1953 年汉斯・韦格纳同建筑师拉斯姆森和玻约森聊天时，谈到就寝时折叠衣服的麻烦，于是就有了这个有趣的设计。为了让它看起来更加轻便，设计师将四条腿改为三条腿。

## Valet Chair

The year of coming out: 1953

Material: oak

In 1953, Hans Wagner, architects Steen Eiler Rasmussen and Bo Bojesen were inspired in their conversation. They were all troubled in folding clothes when they are in school dorm, thus this interesting design came up. To make it more light and handy, designers changed the four chair legs to three.

### (6) 艾洛・阿尼奥 Eero Aarnio

艾洛・阿尼奥 1932 年出生于芬兰，是波普风格家具设计大师之一。艾洛・阿尼奥的许多作品在全球都有很高的知名度，并获得了诸多工业设计奖项。例如，他在 1963 年设计的球形椅，是以玻璃纤维制成的球形椅子，很快就被大量地制造生产，风靡全球。在阿尼奥的设计中玻璃纤维是他最喜欢使用的材料。其他代表作品还有糖果椅、番茄椅和极富未来感的泡泡椅等。

EeroAarnio was born in 1932 in Finland. Many of his works have very high reputation in the world, and received several industrial design awards. For example, the round Ball Chair he designed in 1963 is made of glass fibre. It was soon produced in large number, and universally popularized. The glass fibre is his favored material. Other representative works such as Pastil Chair, Tomato Chair, and Bubble Chiar which with full of future sense. Aarnio was one of the master of Pop Art.

## 球形椅

发表年份：1963

材料：玻璃纤维

从圆形的球状体中挖出一部分或使它变平，可以形成一个独立的单元座椅，甚至形成一个围合空间，因此被称为“房间中的房间”。它提供了一个平静而私密的空间，可令人保持长时间的放松。

## Ball Chair

The year of coming out: 1963

Materials: glass fibre

This is a chair with independent seat, forming by an enclosed round space, which is empty but with a flat seat. It is called a room in the room ". It provides a quiet and private space. It makes people feel relaxed and wants to stay for a long time.

---

## 泡泡椅

发表年份：1963

材料：透明亚克力、钢管

透明的亚克力加上钢圈和椅垫制成的大型吊椅，摆脱了椅子只能放在地板上的呆板印象。悬吊在天花板上宛如飘浮在空气中的泡泡一般，澄净而透明，未来感十足。

## Bubble Chair

The year of coming out: 1963

Materials: transparent acrylic, steel pipe

This chair with large seat is made of transparent acrylic and steel pipe. It jumped out of the stereotype of that chair should be placed on the ground. The chair hanging on the roof and suspended in the air like a bubble, it is clear and transparent, and with full of future sense.

## 小马椅

发表年份：1973

材料：聚酯冷凝泡沫、羊绒布

小马椅是由柔韧性聚酯冷凝泡沫包在金属骨架外面构成的，椅子的表面材料是流行的丝绒。

## Pony Chair

The year of coming out: 1973

Materials: polyester condensation foam, cashmere fabric

Pony chair is made of flexible polyester condensation foam as the skin to wrap the metal skeleton inside. The material of the chair surface is popular velvet.

---

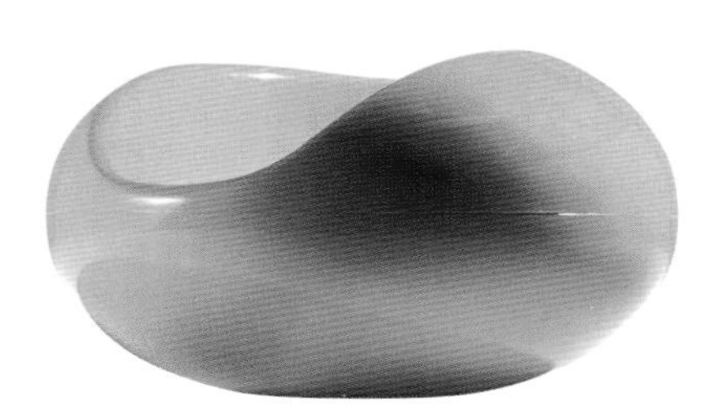

## 糖果椅

发表年份：1968

材料：压模玻璃纤维

阿尼奥在 1968 年设计了糖果椅，这个椅子反映出了美国二十世纪六七十年代自由浪漫的气息。因此他获得了美国工业设计奖。《纽约时代》评价球椅和糖果椅时说：“这是支撑人体的最舒适的一种形态。”

## Pastil Chair

The year of coming out: 1968

Materials: molded glass fibre

Eero designed this chair in 1968. This chair typically represent the romantic atmosphere of 1960s and 1970s in America The design of this chair brought Eero American's Industrial Design Excellence Awards. New York Times made comment on this design that “ This is the most comfortable form of supporting human body. ”

---

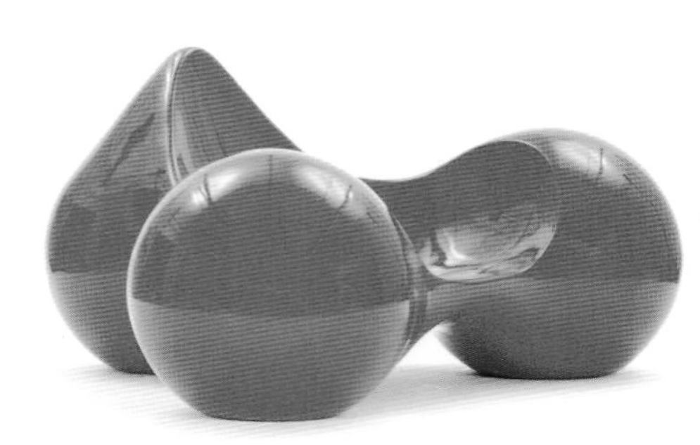

## 番茄椅

发表年份：1971

材料：压模玻璃纤维

番茄椅是由球形椅和糖果椅变化而成的，以流畅的弧线和曲面取代传统椅子的直线和直角，成为椅子造型设计中的经典作品，它由两个圆形的扶手和三角形的靠背构成，独特新颖。

## Tomato Chair

The year of coming out: 1971

Material: molded glass fibre

Tomato chair was transformed from Ball Chair and Pastil chair. The smooth curves and curved surface replaced the straight lines and right angle of the traditional chair. It has become the classical works in the filed of chair designing.Two round chair arms and a triangle back as the structure of this chair, make this design very fresh and unique.

# 3. 北欧家具的主要特征

The Main Characteristics of Nordic Furniture

## (1) 简约化与自然化 Minimal and Naturalized

北欧传统文化之中处处透露着的简洁之美，折射出质朴、真挚、谦逊的人生观以及务实、理性的民族性格。北欧四国的家具造型是符合现代设计的主流风格——简约主义，摒弃烦琐、崇尚简约、重视功能。

Traditional culture of Nordic always reveals the beauty of simplicity, which reflects the simple, sincere, modest life view and pragmatic and rational national character. Nordic furniture design is in line with the modern design of the mainstream style. They are simple and uncomplicated. They advocate concise impression and pay attention to function.

北欧设计是关于自然的美学，源于自然的设计。北欧地区的人们钟爱天然材料，除了偏爱木材以外，皮革、藤、棉布织物等天然材料都被作为设计主材。20 世纪 50 年代以来，随着技术与工艺的进步，北欧也与时俱进，使用镀铬钢管、ABS、玻璃纤维等新人工材料制成经典家具，但整体来说采用天然材料是北欧家具最富有人情味的特征。

Nordic design is about the aesthetics of nature, and it is originated from natural design. Scandinavian people love natural materials. Besides the favored wood,leather, rattan, cotton fabric matter and other natural materials are also used as the main material in designing. Since 1950s, along with the progress of technology and craftwork, Scandinavia also followed the times. They adopt chrome plated steel pipe, ABS, glass fiber and other new artificial materials to make classical furniture, but the natural materials are the most humane features of Nordic furniture.

## (2) 功能与情感结合 Combination of Function and Emotion

纯粹的功能主义是以科技、机械为主导的，忽略传统与文脉，淡薄人情味，工业化的冷漠感给人带来不安，北欧设计提倡关怀人、尊重人、以人为本的设计理念。人情化建筑理论倡导者，芬兰设计师阿尔瓦·阿尔托和机械主义建筑大师勒·柯布西耶的观点相反，他认为自然不是机器，也不应该为建筑的模式服务。同时他还强调：“建筑不应该脱离自然和人类本身，而是应该遵从于人类的发展，这样会使自然与人类更加接近。”

The pure functionalism is mainly based on technologies and mechanics. It ignores the traditions, context, and human nature. The coldness of industrialization brings people anxiety. Scandinavian design advocates loving and respecting human nature, and they stick to the design concept that based on people's needs. The advocator of human nature, Alvar Aalto's architectural theory is in contrast with master architects Le Corbusier's machinery theory. He didn't regard that natural is machine, and should not provided as the service of architectural style. At the same time, he emphasized that architectural should not isolated from nature and human, but following the development of humanity, as it will make people feel closer to nature.

## （3）传统与现代相融合 Fusion of Tradition and Modenity

北欧家具的现代化融合了各自的民族特点和传统风格，而不是割裂现代和传统的关系，因而更容易被北欧接受，具有本民族传统特点和借鉴其他国家传统的北欧现代化家具风靡世界。斯堪的纳维亚地区从 20 世纪 20 年代就开始探索适合自己的现代主义方式，他们既强调现代主义功能至上的原则，同时也强调图案的装饰性，以及传统与自然形态的重要性。丹麦现代设计派鼻祖凯尔·克林特在注重功能前提下，吸收世界各地不同文明的家具设计中的优点，根据特定生存环境，采用传统自然材料制作家具，他的设计理念和方法为北欧设计指明了方向。

Nordic furniture incorporate the traditional and ethic characteristics, instead of cutting the relations of modern and traditions, thus became more easily accepted. The national and traditional characteristic and the reference of other countries' tradition in Scandinavia modern furniture make them swept the world. From 1920s Scandinavia had began to explore a way of modernism that suits them. They stressed on the principle of functionalism but at the same time, also pay attention to the decorative elements of the patterns and the significance of traditions and natural shape. Based on the precondition of stressing on the functionalism, the founder of Danish modernism design Kaare Klint absorbed various characteristics in other civilized countries' furniture. Based on certain living environment, and adopt traditional natural materials to make furniture. His designer concept and approach had showed a direction for Nordic design.

## （4）经济性与生态性 Economic and E-cological

北欧一方面受限于自然资源和能源的匮乏，另一方面受到单纯、朴素的民风的影响，形成了设计上的考究和对资源节约保护的特色。该地区最为丰富的是林业资源，所以设计上普遍采用木材。斯堪的纳维亚四国非常重视生态环境保护工作，这在欧洲也是比较突出的，1994 年颁布了第一个家具生态标准——《木质家具和家具设置的生态标准》，此标准拒绝奢华风的设计，提倡实用经济，设计师们运用创造性设计体现可持续发展的理念，利用设计重新去审视自然界与人类的共生方式，从而营造更美好的生活环境。这一思想理念与设计界 20 世纪 90 年代兴起的“绿色设计”相一致。

Nordic is restricted from short of natural and energy supplies. On the other aspect, they are also influenced by the innocent and simple custom, thus formed the characteristic of conservation in design and energy saving.The region is rich in forest resources, so people generally adopt wood in the design. Four countries in Scandinavia attached great importance to ecological and environmental protection. It is quite outstanding in Europe. In 1994, they promulgated the first ecological standards, the ecological standard of wooden furniture and furniture set up. It rejects luxurious design, and advocates practical and economic design. Designers use creative designs to embody the concept of sustainable development, and apply the design to examine the symbiosis of nature and human, so as to create a more wonderful living environment. This concept coordinates with the "Green Design" which appeared in 1990s.

# 二、纺织品 Textiles

## 1. 北欧织物图案的形成与发展

The Shaping and Development of the Patterns in Nordic Fabrics

由于北欧地区的人们多数信仰基督教，因此带有浓郁宗教题材的图案元素在室内软装中运用得较多。二战后设计风格与流行元素得到前所未有的共享，因此北欧纺织品装饰图案也随之发生了变化，其他国家和民族的时尚元素的引入使得北欧纺织品装饰图案更加丰富和充实，如引入日本漫画图案、英国现代装饰纹样等。

As most of the people in Scandinavia are Christian, the patterns with rich religious elements are applied more in soft decoration. After World War II design style and popular elements have sh aredunprecedentedly,therefore, the patterns in fabrics have changed. The introduction of fashion elements in other nations made Scandinavian textile decorative patterns more abundant and diverse, such as the introduction of Japanese cartoon pattern and the modern decorative patterns.

## 2. 北欧织物软装饰中的图案元素

The Elements of the Nordic Textiles Patterns

在现代室内纺织设计中融入传统图案是一个发展趋势，传统的图腾、桦树叶、花卉图案是北欧纺织图案永恒的主题。北欧的织物软装图案中也有可爱的动物元素，例如具有浓郁北欧风情的麋鹿图案，充满民族风情的小马图案等。

北欧的织物软装饰常用图案从时间上来分，可以分为传统图案和现代图案两大部分，从图案的内容来分可以分为自然图案、动物图案、几何图案、卡通图案、速写图案、混合图案等。

It is a trend to make traditional patterns in modern interior textiles. The traditional totem, birch leaf, floral pattern are always the theme elements in Nordic textiles. Besides, there are also lovely animals pattern in the textiles, such as elk, which has strong Nordic flavor and pony pillow, also full of ethnic customs.

The common patterns on Scandinavian textiles can be divided into traditional patterns and modern patterns. The content of the patterns can be classified into natural patterns, animal patterns, geometrical patterns, cartoons, sketches and mixed patterns etc.

瑞典室内传统纺织品

The traditionalstyle of Swedish interior textiles

## 自然图案

北欧纺织品种使用最多的图案，树木、花朵给寒冷的北欧家居带来生机和活力。

## Natural Patterns

are the patterns that mostly applied on Nordic textiles design. Trees and flowers bring vitality and energy to the concise Scandinavian furniture.

## 卡通图案

充满童趣，适合用于装点儿童房。

## Cartoon Patterns

are full of childish delights. They match well with children' s room.

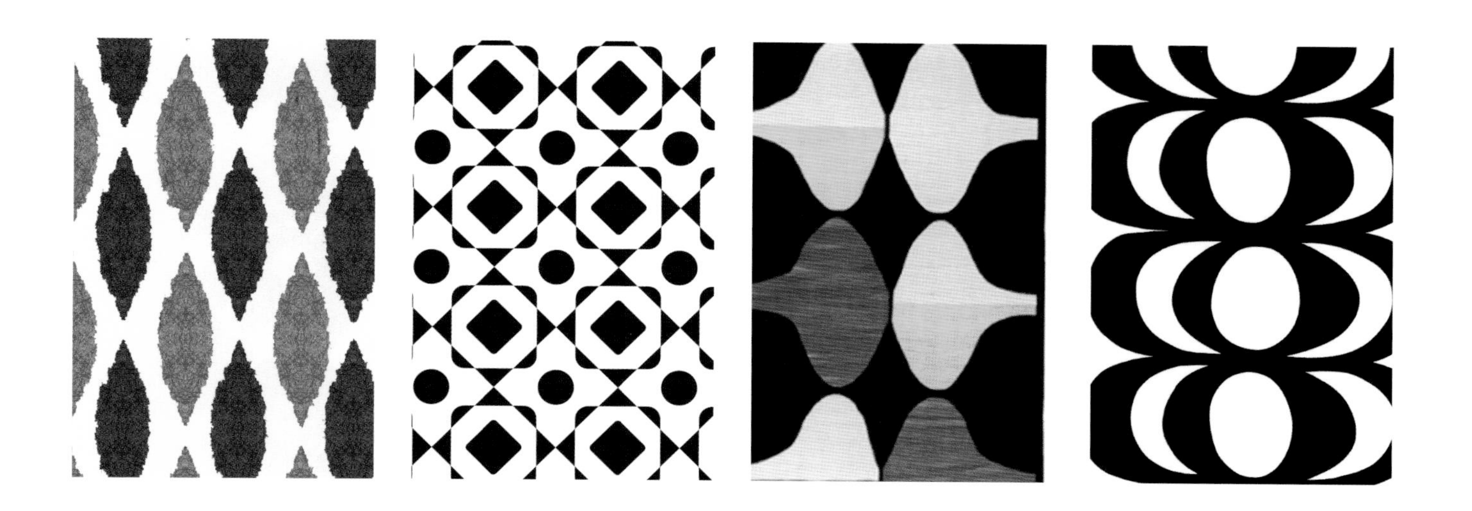

## 几何图案

简约、时尚、理性、完全吻合北欧风格的设计理念。

## Geometrical Patterns

are t concise, fashionable, and rational. They match perfectly with reubdeer design concept.

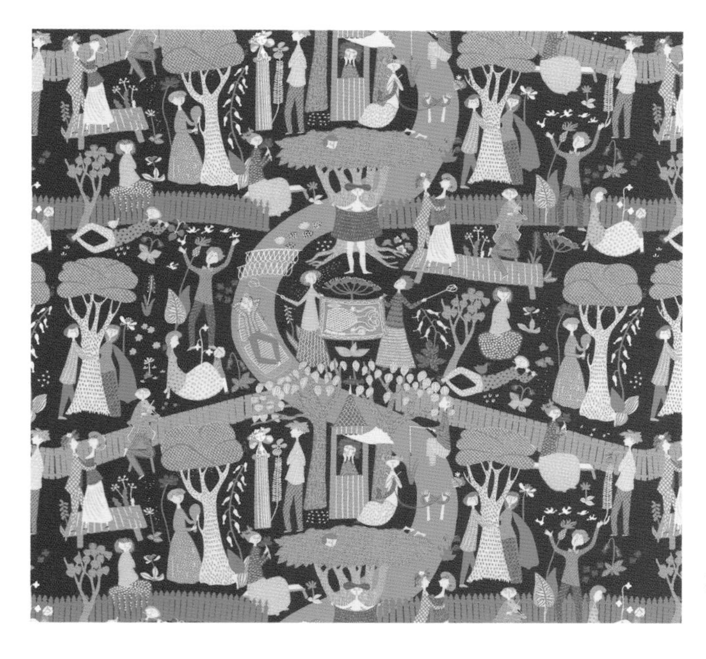

斯蒂格·林德伯格设计

Designed by Stig Lindberg

## 传统图案

以自然和民间生活为主题，表现出浓郁的北欧民族风。

## Traditional Patterns

the natural and folk life as the theme, showing a strong Nordic national style

## 3. 北欧织物软装饰的特色

The Characteristics in Nordic Textiles Decorations

织物是北欧家居的主角，设计师把生活中常见的动物、植物，以一种图案化的、简洁的方式表达出来。在寒冷漫长的冬天，缺乏阳光的日子使得北欧织物绝不能缺少活泼的几何图案，多种多样的织物图案展现了北欧设计师从抽象艺术和传统图案中获得的想象力。太复杂、古典、异国情调的图案都不适合北欧风格家居，北欧的织物设计总体来看倾向图案简洁、淳朴自然、色彩鲜艳明快的特点。

Fabric is the crucial elements in Scandinavian house decoration. Designers take our familiar living elements, such as animals, plants, to express simply in the patterns. In the long winter nights in Scandinavia, lively patterns are indispensiblein their textiles. The variety of patterns manifests the imaginations which Scandinavia designers inspired from traditions and abstract art. The complicated, classical and exotic plants are not match with Scandinavia furniture. Scandinavia textile decorations are mostly in simple and natural patterns, and bright colors.

棉麻因为质感平易近人且容易打理，自然被认为是最质朴的设计材料。人们除了用棉麻织物制作抱枕、沙发和窗帘以外，还用它们制作灯罩、布帘隔断和装饰品等。

Cotton linens are easy to clean, so it is regard as the simplest decorating materials. People use cotton linens to make pillows, sofa and curtain, and other ornaments such as lampshade and Partitions etc.

地毯在寒冷的北欧冬季能营造出一种舒适、温暖的家居氛围。造型简约、色彩鲜明的北欧风格地毯比较流行，简单的设计不等于朴素，良好的色彩搭配加上有内涵的设计，一样也能呈现出独具特色的北欧风情。

Carpet can create a comfortable and warm ambience in the cold winter of Northern Europe. The clean and fresh Scandinavian style's carpet are very popular. The simple designs that match with various elements delivers a low profile luxury.

# 三、其他装饰品 Other Ornaments

## 1. 木制装饰品
Wooden Ornaments

木材在北欧风格设计中占据着极为重要的位置，从闻名世界的北欧家具到各种木器小摆件，北欧风格的室内空间木器是必不可少的角色。选用的木材也可以是粗犷而简约的，并不需要精工细磨，没有经过油漆上色的原木让北欧风格显得更加原生态。常用的木质装饰品有北欧风情的木质麋鹿头、达拉木马和宜家吉特达木人偶等。

Wood always takes the position of the soul in the Nordic style. From the world famous Scandinavia furniture to a variety of wood decoration, wood in Scandinavia interior is an indispensable role.The selection of wood can be crude and simple, and does not need the fine grinding. Logs without painting can present the most original side to the Scandinavian ecological condition. The commonly used wooden decorations are Scandinavian style wooden moose head, Dallas Trojans and IKEA Gestal puppets.

北欧风格麋鹿墙饰
Nordic moose head

瑞典达拉木马摆件
Swedish Dallas Trojans

## 2. 陶瓷、玻璃饰品

Potteries and Glass Ornaments

陶瓷、玻璃、铁艺等经常作为装饰品或作为绿色植物的容器，出现在北欧风格的室内，不管是装饰品还是容器往往都保留了材质的原始质感，体现了北欧人对传统手工艺和天然材料的喜爱。其中玻璃装饰品最常见的就是形态简洁现代的花瓶。1936年阿尔瓦·阿尔托负责赫尔辛基甘蓝叶餐厅的室内设计，并亲自设计了一款花瓶作为装饰品，以芬兰星罗密布的湖泊作为设计灵感来源，完全不同于传统对称古板的玻璃器皿的设计形态。阿尔托不仅设计了许多出色的建筑，也为玻璃器皿制造业留下了经典杰作。尽管阿尔托花瓶已经问世多年，但应用在现代室内环境中它仍然很时尚、很现代。

瑞典玻璃制品在1925年的巴黎国际博览会上赢得了极大的关注，并一举打进了美国市场。但最值得一提的是丹麦的工业设计，由汉宁森（Poul Henningsen）设计的照明灯具在博览会上广获好评，在该届博览会上被誉为是唯一能与柯布西耶的“新精神馆”相媲美的优秀作品，并获得金牌。这种灯具后来逐渐发展成了著名的PH系列灯具，至今畅销不衰。PH系列灯具既有极高的美学价值，又基于照明技术的科学原理，而不是多余附加的装饰，现实室内空间中使用简洁且效果极好，这正体现了北欧工业设计的特色。

Potteries, glass and ironwork are often adopted in Nordic style interiors or the container of plants. No matter ornaments or containers will remain their original texture, it manifests Nordics' favor for traditional handcraft and natural materials. Among those materials, glass is the commonest to make vase, and often present in concise and modern shape. In 1936, when Alvar Aalto was in charge of the interior design of Ravintola Savoy, he designed a vase that based on the inspiration from the densely distributed lakes in Finland. It is completely different from the traditional symmetrical vase shape.Alvar Aalto was not only talented in Architectural design, but also left many classical masterpieces. Although Aalto vase had come out for many years, but it is still fashionable and modern in todays' interior design.

Swedish glass products had won great attention in the 1925 Paris International Fair, and in one fell swoop into the United States market.The most worth mentioning is the Danish industrial design. Henningsen's lighting design had won wide praise in the fair. It is reputed as the only excellent workthat comparable to Corbusier's Pavillon de L'Esprit Nouveau, and won the gold metal. This lamp has gradually developed into famous PH series lamps, and has been sold well. PH lamps series not only has a high aesthetic principle, but also based on the scientific principle of the lighting technology, rather than the extra decorative. It matches perfectly with simple interior spaces, which reflects the characteristics of Nordic industrial design.

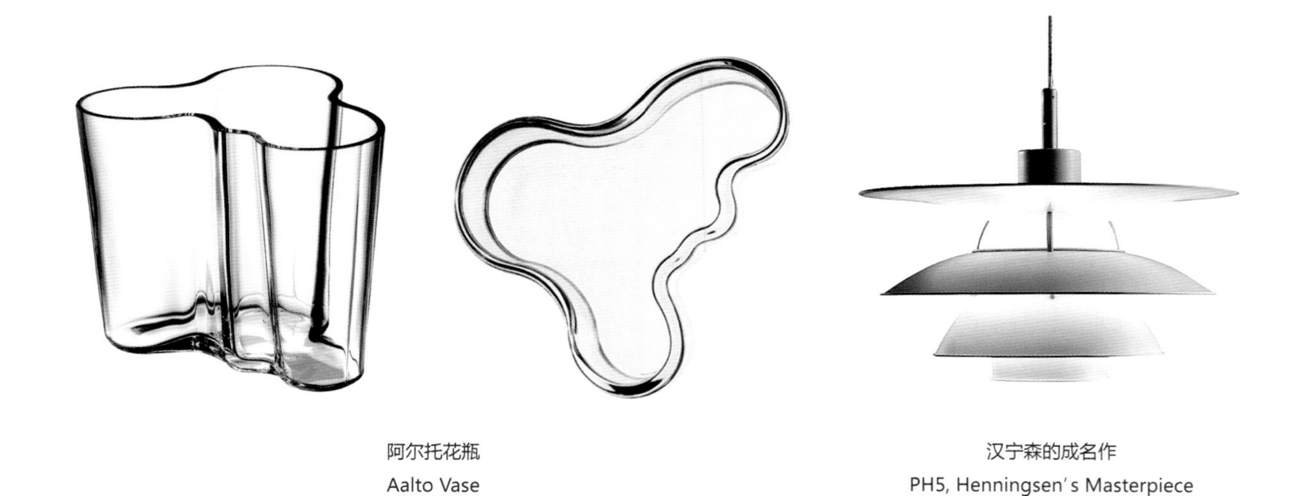

阿尔托花瓶
Aalto Vase

汉宁森的成名作
PH5, Henningsen's Masterpiece

## 3. 装饰绘画

## Decorative Painting

装饰绘画也是北欧室内设计中最常见的装饰元素。装饰绘画主要配合整体设计作为点缀，有时也会成为视觉的中心。北欧在绘画题材方面多以自然元素为主，以植物为题材，如花卉、树木等；以动物为题材，如猫头鹰、熊、飞鸟等；以自然现象为题材，如星河、北极光等。其绘画手法不一，常见的有写实、抽象、涂鸦、卡通等多种技法，在室内与其他装饰品一起营造一种轻松、自然的空间氛围。

Decorative painting is the commonest decorative elements in the Nordic interior design.Decorative painting mainly applied as ornament to match with the overall design sense, but sometimes it also can be the center of the visual.Scandinavian paintings have multiple natural elements, such as flowers, trees and other plants; animal as the themes, such as owls, bears, birds and other subjects; natural phenomena theme, such as the starship, northern light, etc.The painting techniques are different, it is common to have a variety of techniques, such as realism, abstract, graffiti, cartoon, etc., in the interior and other decorations together create a relaxed, natural atmosphere.

# 四、色彩 Colors

北欧风格的室内色彩呈现明亮、艳丽的特征，整体倾向于干净明快的浅色系。通常以白色为基调，配合浅木色，局部使用高明度、高纯度的色彩加以点缀。形成这种室内色彩的原因也与北欧的自然地理因素和北欧人民的性格有关。北欧地区因为处于北极圈附近，气候比较寒冷而且缺少日照，有些地方还会出现长达半年的“极夜”气候。虽然自然色彩不比热带、亚热带地区的丰富多彩，但北欧人的内心充满热情与活力。因此大面积的沉闷的颜色不适合在北欧居室空间中运用，北欧人通常会用高纯度的颜色装饰局部，例如清浅的蓝色、荧光黄、明亮红等，营造室内空间的活泼气氛，使人们在居室环境中不会产生沉闷和厌烦的情绪。对于北欧的室内色彩来说首先要控制好色彩的面积平衡，避免室内出现灰暗的主色调，其次在家居的装饰配色上，用高纯度颜色打破北欧常年被雪覆盖的单一白色的单调感，用艳丽的色彩加以点缀纯净的室内空间来弥补自然色彩的单一性。

Nordic style indoor color favored in bright, colorful characteristics, the overall colors tend to be clean and bright, no messy extra colors. White is commonly adopted as the keynote, with light color and wood based.It often matches with high brightness in partial, and high purity color as extra decoration. The reason for the formation of Scandinavia interior color is related to the natural geographical factors and the people's character. The Scandinavia region nears the Arctic Circle. The climate is cold and lack of sunshine. In some places, still appear polar night climate for up to six months.The natural colors are not colorful compared with tropical and subtropical regions, but the people have a fiery and vibrant heart. Therefore large areas of dull colors are not suitable for Scandinavia interior spaces, usually with high purity color anddetail decoration, such as shallow blue and fluorescent yellow, bright red.Creating lively atmosphere in interior spaces can avoid of people being stuffy and boring.For Scandinavia indoor color, the first step is to control the colors and the area of the balance. Gray colors should be avoided to applied as the keynote. Secondly, in selection of furniture, high purity colors can be used to cover the white color, and with bright colors to decorate the interior space to make up for the single natural color.

**纯白的墙面能让空间看起来更加宽敞明亮，但是如果没有出彩的配饰，就显得平淡无奇，将一面墙刷成彩色，空间立刻就变得有层次感了。**

White walls can make the space more bright and wide, but if without colorful accessories, it will looks plain. Paint one wall with color will immediately highlight the whole space.

设计师用明快的色彩装饰以白色为主色调的空间，其中黄色扮演着重要的角色。

Designers use bright colors and white as key note in this space, among them, the yellow played an important role.

地板、书柜、矮柜、沙发等采用同一个色系；营造了一个统一和谐的家庭氛围，大地色系也是一个比较容易协调的色系。

A unified color scheme is applied to floor, bookshelf and low cabinet, making an overall harmony  family atmosphere. Earth-toned color is also relatively easy to coordinate the color system.

室内纺织品由于使用面积较大，其色彩基本决定了整个家居空间的色彩主调。北欧室内织物装饰通常选用面积较大且鲜艳的颜色，而且以暖色居多，织物软装饰的色彩和图案无论在白天的阳光下还是夜晚的灯光下，都易于营造温馨感，反映出了自然惬意的北欧生活方式，从而可以在寒冷的冬季获得更多的温暖感受，这也符合人们心理的需求。例如白色墙壁与窗帘，配以暖色系的织物，营造出室内的明亮暖色调，局部可增加高纯度的靠枕等陈设品作为对比，以打破色彩的单调。

The textiles take up a large area of the indoor space, therefore their colors basically determined the while keynote of the home decorating. Scandinavia indoor textiles mainly adopt bright and fresh colors with large area, and mainly in warm colors.The colors and patterns of the fabrics can make people feel warm and comfortable no matter in the daytime or in the night. It reflects the casual lifestyle in Scandinavia, and manifests their needs of more warmness in the cold winter. Such as the white wall and curtains are matched with warm colors fabrics to create a bright and warm keynote in interiors. Partially can decorated with high purity pillows or other displaying tomake a contract, and break the single color thus retain a visually esthetics.

Ⅲ

第 3 章 Chapter 3

# 提升品位的北欧设计

ENHANCING NORTHERN EUROPEAN INTERIOR DESIGN

# 北欧自然风格
# Natural Nordic Design

## 风格概述 Style Description

北欧自然风格是指崇尚自然、与自然环境相融合的、质朴的室内设计风格。它所呈现出来的是非常接近自然的原生态的美感，以房屋自身结构构件作为装饰，一切材质都尽量保留自然的肌理和色泽，重视传统手工，并强调与户外大自然的融合。

Natural Nordic style refers to the natural environment and the integration of the simple interior design style. It presents an original beauty which is very close to the natural ecology. Without superfluous decoration, all materials remain their natural texture and colors. It stress on traditional handcraft and harmony with the outdoor nature.

## 细节设计 Detail Designing

北欧自然风格在材质上，一般选用自然的原木、天然的石材和棉麻的织物等。保留自然材料的质朴纹理，其形态和外表越自然越好，色彩上也基本上保留原始物料的颜色。在空间处理上重视采光与整体空间的通透性，并且尽量将一些室外的自然景物引入到室内，达到一种室内外情景交融的感觉。在软装饰上常有藤制品、绿色盆栽、瓷器、陶器等摆设。布艺则多采用以植物和传统元素等为图案的棉麻织物。

Nordic style generally use natural wood and original stone. All materials remain the natural texture. Their shape and colors are all prevented their natural status. The light should be most transparent in spaces, and try to lead the outdoor landscapes into indoor to create a beautiful scenery. In soft decorations, there are rattan products, plants, pottery and porcelain. The fabrics with plants and traditional patterns and in cotton fabric are mostly applied.

## 装饰技巧 Design Techniques

▶ 色彩既可以是清新自然的原木色，也可以是充满民族风情的浓墨重彩，例如传统的黑、白、红。
Colors can be both fresh color and wooden color, and also can be with full of splendid ethnic customs, such as the traditional black, white and red.

▶ 常用的装饰材料有松木、白蜡木、橡木、石材、藤编、铁皮、粗麻织物等。
The commonly used decorative materials are pine, ash, stone, rattan, iron, hemp, fabric etc.

▶ 未经处理的砖石墙面，或者原木的吊顶，都体现了空间的质朴感觉，也是北欧自然风格中经常运用的一些装饰手法。
The walls and bricks without any treatment and wood ceiling reflect the rustic feeling of the space. It is also commonly used as the decorative treatment in natural Nordic style.

▶ 陈设与装饰力求淳朴、自然，可以选用造型简洁质朴的陶艺饰品、传统手工图案的纺织品、原木或者藤编的小饰品、以北欧风景为主题的装饰画等。
Decorations and display strive to be simple and natural. Ceramic accessories, fabrics with traditional handcraft patterns, wood and other rattan accessories and decorative panting with Nordic landscapes can be used indoor.

# 伯利恒山之居

# Bethlehem Hill Transitional

项目地点：美国，新墨西哥州，马德里镇 ❉ Location: Madrid, New Mexico, America

项目面积：279 平方米 ❉ Project Size: 279 m$^2$

设计公司：French & French Interiors ❉ Design Studio: French & French Interiors

摄影：Bill Stengel、Marc Risik ❉ Photographer: Bill Stengel, Marc Risik

本案坐落于一座狭长的平顶山上，可以远眺圣达菲的景色以及整个马德里镇，是一对夫妇自己建造的房子，并在入住期间进行了装修。担任装潢师的 Heather French 和担任建筑师的 Matt French 搭配得天衣无缝。在 Heather 最爱的物件当中，有一些是年轻时热爱设计的她收藏或者买下来的，而 Matt 则从 12 岁开始就从事建筑方面的工作，俩人一直以来都热衷于手工制作。

The house sits atop a finger mesa and has views of Santa Fe and the village of Madrid.The self-taught couple built and decorated this house themselves while living in it. Decorator Heather French and her builder husband, Matt, are well matched. Some of Heather's most beloved pieces are things she saved up for and bought as a design-obsessed teenager. Matt has been building since he was 12. And both have been avid DIYers as long as they can remember.

这个空间原先是房子主体，如今用来作为房子的入口。打开推拉门，就可以看到一块约 74 平方米的木质平台，这也是孩子们最喜欢待的地方，因为 Isla（房主的女儿）的房间和娱乐室就在旁边。一个室内秋千给房子平添了几分乐趣，此外还有一张桌子可供假日宾客聚会时使用，而全家人也很喜欢在这吃早餐。

This space that now serves as the entry used to be the entire house. The sliding-glass doors lead to an 74m$^2$ deck. This entry space is most often used by kids these days, as Isla's room and the playroom are right off it.  An indoor swing adds to the fun. A table serves overflow guests during holiday gatherings, and the family often enjoys eating breakfast here.

Yogurt 是 Isla 的猫，以她最喜欢的食物（yogurt 意即“酸奶”）命名。椅子是二手店淘来的；Heather 用 Nobilis 生产的条纹布给这些椅子重新上了椅面。夫妻两人喜欢收集复古风格的黄铜饰品，这些制品在房子中随处可见。

Yogurt is Isla's cat, named after her favorite food. The chairs were thrift store finds; Heather reupholstered the seats in a striped fabric by Nobilis. The couple loves to collect vintage brass accents, seen throughout the house.

除了收集古董、工作与DIY外，夫妻两人还喜欢收集当地艺术家和工匠的艺术品和家具。这张皮革饰面的桌案出自当地工匠Raymond Linam之手，桌子的波浪式钉头图案与夫妻两人的黄铜藏品搭配和谐。照片中是当时出席俩人婚礼的所有宾客。

In addition to antiquing, thrift and DIY projects, the couple loves to collect art and furniture pieces from local artists and craftspeople. Local craftsman Raymond Linam made the leather-wrapped console table. Its nailhead scalloped pattern picks up on the couple's brass collections. The photograph includes everyone who attended the couple's wedding.

走到大厅，能看见一个讨喜的储藏柜，镶嵌着精致的黄铜。这个柜子是他们的一个亲戚到印度旅游时给他们买的。柜子上摆着当地艺术家Mat Crimmins制作的浇铸式青铜雕塑。French夫妇收藏了Mat Crimmins的这个雕塑和他另外的几件作品，还收藏了当地艺术家Nigel Conway的几件佳作，包括柜子上方的那幅挂画。

Down the hall, a favorite cabinet with an intricate brass inlay was given to the couple from a family member who brought it back from a trip to India. On it sits a cast bronze sculpture by local artist Mat Crimmins. The couple owns several of his pieces as well as several by local artist Nigel Conway, who made the painting over the cabinet.

房子的这一块是一个很开放的空间。靠左居中的位置放着一张大理石台面办公桌，靠右是厨房，空间的末端则是客厅。整座房子的天花板均由当地的冷杉和松木制作而成，其中大部分的木材都是在森林大火之后进行清理工作时搜集来的，夫妻两人买下这些粗糙的木料之后就自己动手切割打磨，墙面采用的是威尼斯石膏。

This part of the home is one wide-open space and includes a marble-topped command central to the left and the kitchen to the right. The end of the space is the living room. The ceilings throughout the home are locally sourced fir and pine, most of which was harvested during clearing after forest fires. The couple bought the lumber rough-cut and sanded it themselves. The walls are Venetian plaster.

客厅的另外一个焦点就是 Nigel Conway 所绘的一幅挂画，画的两侧为 Matt 自制的两盏灯，他将木头烧焦，并控制好烧焦的程度，之后将这些烧焦的木头进行封蜡，以防剐蹭。榻床式的咖啡桌上摆放着一尊 Mat Crimmins 制作的青铜象骨架。夫妻两人都喜欢收集新墨西哥的艺术品，这里的家具几乎都是从古董店和二手店淘来的。（多彩抱枕来自 Anthropologie；大一点的枕套来自 Pindler & Pindler）

The focal point in the living room is another painting by Nigel Conway, flanked by two lamps handmade by Matt. The coffee table is an opium bed. The bronze elephant skeleton on the coffee table is by Mat Crimmins. The couple enjoys collecting New Mexico artwork. The yellow chair is the first piece of furniture Heather bought as a teenager.（Colorful throw pillows: Anthropologie; larger pillow fabric: Pindler & Pindler）

厨房是开放式的， 隔板的金属托架以及中岛的支架均由当地金属制造工 Mike Forloney 制作，隔板的材料为松木。大面积的中岛是聚会空间，同时也是一个很实用的操作台，这种设计现下很流行。中岛台面下边的空间除了做垃圾回收之外，还存放着各种壶和锅。此外，Isla 也有一个存放自己宝贝的柜子。夫妻两人把香草、调味品和烹饪时用到的物品以及茶放在较低的隔板上，以方便拿取。复古风格的爆米花机等一些他们至爱的物件则让迷人的摆设锦上添花。吊灯选用的是简约透明玻璃材质吊灯，这样就不会与窗外的景色相争。（炊具、通气罩和冰箱来自 Viking，洗碗机抽屉来自 DCS）

The whole kitchen is surrounded by open shelves. The metal brackets as well as the island's base were made by local metalworker Mike Forloney. The shelves are pine The large island is a popular gathering space, as well as a useful work surface. Below, they store their recycling station and pots and pans, and Isla has a cabinet for her favorite things. Herbs, spices and other things the couple uses while cooking, as well as tea, are conveniently located on the lower shelves. Some of their favorite items, such as the vintage popcorn popper (top left), contribute to a beautiful display.The couple chose simple clear-glass pendants so as not to compete with the view out the window. (Range, vent hood and refrigerator, Viking; dishwasher drawers: DCS)

主卧室一侧的床头柜是与客厅相同的镶嵌着黄铜的储藏柜，另一侧则是一张复古样式的桌子，床头的上方是两幅产自印度尼西亚的挂画，由手缝的布料制成，以竹子做边框。床尾的脚凳面是用莎丽布料编织而成，而镜子则是 Heather 年轻时淘到的一件古董。

In Heather and Matt's bedroom, another brass inlay cabinet from India serves as a nightstand; an antique desk serves as the other. This chest was given to them by a well-traveled family friend, who also brought them the two works over the bed, hand-blocked pieces of fabric framed in bamboo from Indonesia. The bench at the foot of the bed is woven from saris. The mirror is another antique that Heather picked up when she was very young.

主卧床正对着的是一张复古式的橙色书桌。墙上的装饰挂毯为反面放置的摩洛哥婚毯，他们都喜欢毯子背面这些精致的编织图案。桌子上摆放的是一套三只的黄铜鸟以及一个满是烟斗的鱼缸。虽然他们都不吸烟，但是在一个车库甩卖会上看到这个装满木制烟斗的鱼缸时，一下子就喜欢上了它，因为烟斗的木料漂亮极了。

The couple's bed faces with this desk. The textile is a Moroccan wedding blanket turned backward. They both loved all of the intricate weaving you can see on the back side. A trio of brass birds and a fishbowl full of tobacco pipes decorate the desk. Neither one of them smokes, but they found this bowl full of burled wood pipes at a garage sale and they really liked it — the wood on the pipes is beautiful.

主人房浴室的 Thomas Pell 挂画是在一次房产买卖中淘来的，吊灯的材质是银色玻璃，浴缸来自 Vintage Tub & Bath，盥洗台来自 Signature Hardware，水龙头来自 Hudson Reed。通过镜子的反射，你可以看到淋浴隔间，位于图片所没有呈现的房间右侧。在浴缸的右边是一个独立的卫生间，左边的内置门则通向夫妻两人的贮藏室。

The couple designed their master bathroom around the freestanding bathtub. The Carrara marble-topped sink console was ready-made. The Thomas Pell artwork was an estate sale find. The pendants are mercury glass. You can see the reflection of the shower stall in the mirror; it is located just out of view on the right. To the right of the tub is a separate toilet closet. The pocket door on the left leads to the couple's closet.

为了拍照，他们对女儿 Isla 的房间稍做布置。Isla 知道如何摆放所有的物件，而且大部分时间她都能做到物归其位，她喜欢去探知要找的东西具体放在哪里。孩子的房间应该是爱和功能性的结合，而非为了设计而设计。在右边的墙面上，Heather 贴上了一些夜光星星，并把它们设计成流星破空的形状。 夫妻两人把 Isla 的浴室设计成一个能与她共同成长的浴室。如果 Isla 厌倦了粉红色褶边的淋浴帘，可以很轻易地将它换掉。手铸抛光的混凝土盥洗台、浴缸四周以及一些边边角角均由 Matt 手工制作。淋浴帘子和大象挂钩来自 Anthropologie，长颈鹿水彩画来自 Matallo Gallery。

Heather swears that Isla's room is only slightly staged for the photo. Isla knows where everything goes, and she really does keep everything in its place most of the time. She likes knowing exactly where to find what she's looking for. A child's room should be about pure love and function as opposed to just design. To the right, Heather arranged the glow-in-the-dark stars in a shooting star shape. The couple designed Isla's bathroom to grow with her. When she gets tired of a pink ruffled shower curtain, it can be easily switched. Matt made the hand-cast polished concrete counter and tub surround as well as the vanity. Shower curtain and elephant hook: Anthropologie; giraffe watercolor: via Matallo Gallery.

# 少男少女套房
# Jack and Jill Suite

项目地点：美国，圣达菲 ❄ Location: Santa Fe, USA

设计师：Heather French, Matt French ❄ Designer: Heather French, Matt French

摄影：Bill Stengel, Christopher Martinez ❄ Photographer: Bill Stengel, Christopher Martinez

设计公司：French and French Interiors ❄ Design Studio: French and French Interiors

青铜像：Mat Crimmins ❄ Bronze Figures: Mat Crimmins

艺术品：Lori Swartz ❄ Art Work: Lori Swartz

这个大空间的儿童卧室是由 French and French Interiors 为 2014 年“圣达菲展示屋”而设计的，主题为“远古的未来”。作为一个慈善组织，圣达菲展示屋专门为儿童艺术项目募集资金。2014 年，该组织共募得善款 5.4 万美元。在活动中会分派给每位设计师一个房间，根据主题进行设计。分派给 French and French Interiors 的夫妻档的是少男少女套房。该房间空间很大，需要一个庞大奢华的设计，因为从阳台可以俯瞰一个直升机停机坪、奥运会标准规模的泳池以及圣达菲的迷人景色。

This spacious child's bedroom was designed by French and French Interiors for ShowHouse Santa Fe 2014, with the theme "ancient future." ShowHouse Santa Fe is a charity organization that raises money for children's art programs. They raised around 54 thousand dollars for charity in 2014! Each designer is assigned a room in the show home, to decorate according to the theme. The husband and wife team of French and French Interiors were assigned a Jack and Jill suite. The spacious room calls for a large, luxurious design, as the balcony overlooks a helipad and the olympic-sized pool, in addition to a fabulous view of Santa Fe.

设计师是为 5 岁儿童而设计的这个房间，而整个房间最受老少欢迎的地方在于把滑梯与壁橱结合在一起。从滑梯下方进入之后，小孩子可以通过一个矮梯爬到一个读书角和平台。从里边出来的时候，只需从滑梯上滑下来就可以。

French and French designed the room with a five-year-old in mind, and the most popular feature of the room was the slide built into the closet. The child enters beneath the slide and climbs a short ladder to reach a reading nook and platform. To exit, the child slides down the slide.

床头板与被单的条纹为同一款式。床左侧的柜是充满工业化气息的金属风格，而右侧的床头柜则是天然原木材料。青铜材质的犀牛和大象骨架分别摆放在两侧的床头柜上。床上摆放了质地柔软并带有奶油色流苏的毛毯和数个枕头。一只可爱的复古式泰迪熊端坐在枕头中间。

The headboard behind the bed is striped in the same pattern as the bedskirt. The dresser to the left of the bed is in an metallic industrial-style, while the right nightstand is a natural wood. Bronze skeletal figures of a rhino and an elephant sit on the two surfaces on either side of the bed. There is a soft blanket with tassels in cream and the multitude of pillows. An adorable Teddy Bear in an old-fashioned style sits amongst the pillows.

金属材质的床头柜有着钉头镶边，这种镶边赋予了它工业化的气息。从这张照片，我们可以看到青铜犀牛雕塑及其背后的装饰画的细节。

The metallic dresser has a nail head trim that gives the piece an industrial look. From this view we can see the detail on the bronze rhino sculpture and the paintings behind it.

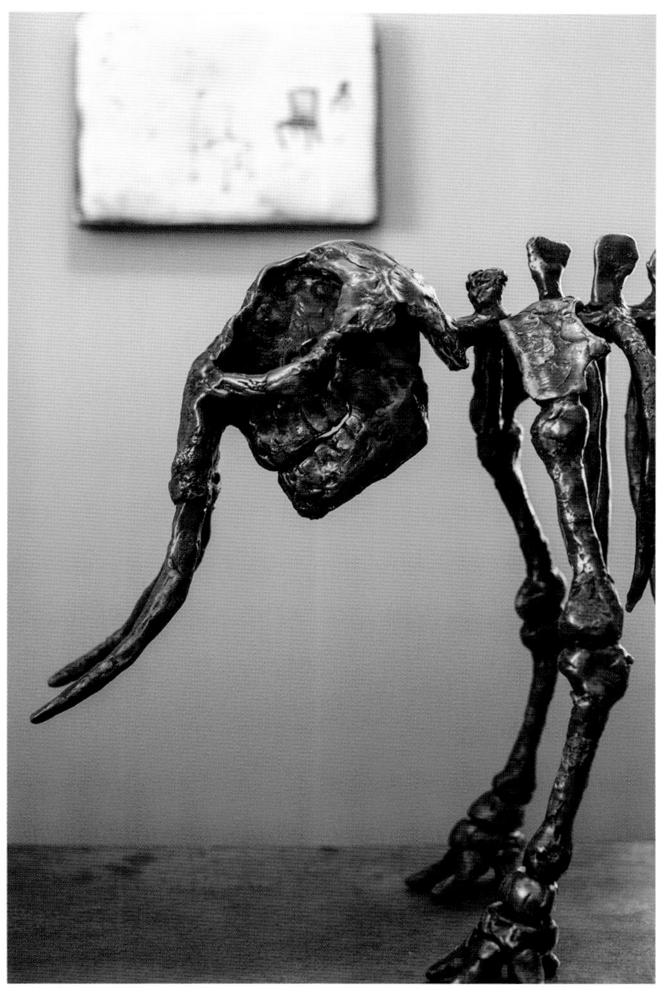

以艺术画为背景的浇铸式青铜大象雕塑。

The elephant sculpture in cast bronze, with artwork in the background.

颇具现代感的白色游戏桌旁放着一把柔软而有质感的懒人椅。一大张编织地毯和一条柔软条纹毛毯共同营造出一个舒心的小角落。机器人挂画是 Nigel Conway 的手笔。

The white modern play table sits near a very soft and highly texturized reading chair. The large weave of the fabric, coupled with a soft, striped blanket, makes for a cuddly corner nook. Robot painting by Nigel Conway.

游戏桌上方的灯具采用青铜作为装饰，显得更为现代化。

The light fixture above the play table is much more modern, with a bronze finish.

游戏桌上简易的木质和纸质玩具。

The simple wooden and paper toys on the play table.

粗麻布枕头以及被当作边几用的编织团。

The burlap billow and textured ball that acts as a side table.

阅读角上唯一的照明设备是用石头和金属丝做成的，上面有小洞，好像有小鸟栖息在里边。

This unique light fixture above the reading nook appears to be stone and wire, with small nooks, as if birds were to be nesting in them.

# 55 平方米迷你住宅

# The 55 $m^2$ Apartment

项目地点：以色列，特拉维夫 ❄ Location: Tel Aviv, Israel

项目面积：55 平方米 ❄ Project Size: 55 $m^2$

设计师：Raanan Stern ❄ Designer: Raanan Stern

摄影：Gidon Levin，181° ❄ Photographer: Gidon Levin, 181 degrees

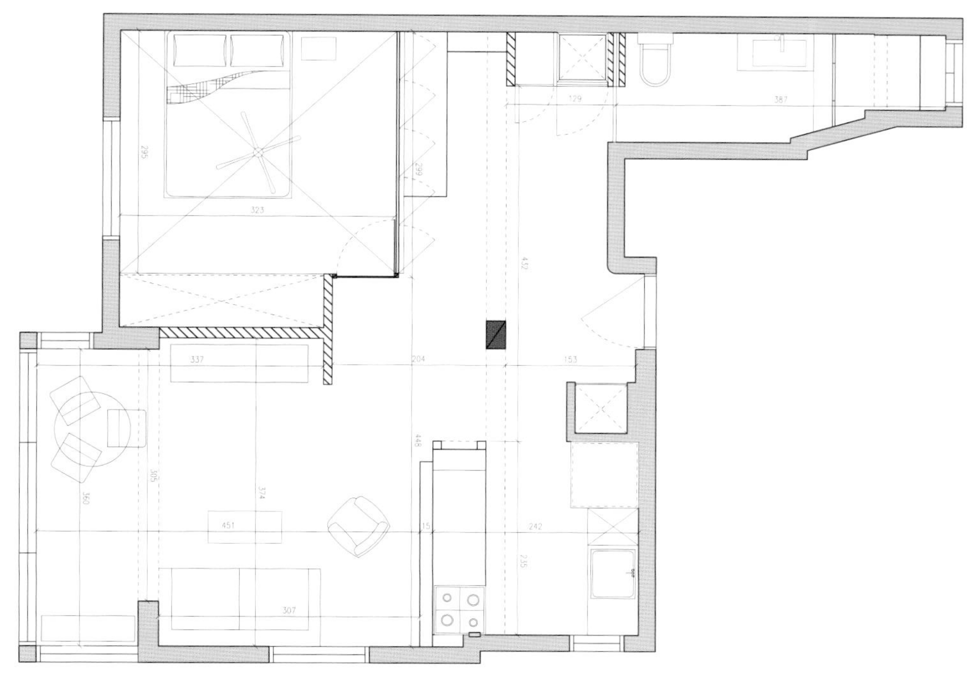

**平面布置图**
Apartment Plan

本案为特拉维夫中心一栋有 70 年楼龄的旧居民楼，由建筑师 Raanan Stern 设计，其再整修的重点是房屋里三个指定方向的通风和光线。原先的公寓在被拆除之后，只保留下一根构造柱和一排窗户。天花板上的风扇赶走了炎热的特拉维夫空气中的潮湿，从而实现了空气流通。原先的浴室被改造成了一个正对客厅的开放式厨房。一面很大的玻璃隔墙镶嵌在薄薄的白色金属框之中，形成一堵透明的墙，将光线从西面引入室内，光线透过卧室，直到工作区。钢化玻璃挡住了强光，并将卧室隐藏在公寓中。在夜幕降临前，这堵玻璃墙能保证不需要人工照明的情况下的正常工作。

The key to the renovation of the apartment, located in a 70 year old neighborhood building in central Tel Aviv, was natural light. The 55 $m^2$ apartment was designed by architect Raanan Stern and the plans functionality was based on the three given directions of air and light. The original apartment was dismantled, leaving one constructive column and window placements. The natural and artificial circulation of air was achieved with ceiling fans mitigating the humidity and heat through the hot and dry Tel-Aviv weather. The old bathroom was opened and turned into an open kitchen facing the main living area. A large partition glass wall with a slim white metal profile creates a transparent wall that brings light in from the west facade. The light enters through the bedroom and reaches the work space. The use of tempered glass prevents glare and hides the bedroom in the apartment. The glass wall allows work to continue without artificial light until evening hours.

KANDINSKY

KANDINSKY

白帆布似的墙面上多种材料与色彩的使用赋予了该公寓鲜明的特征。荷兰叉骨式镶木地板可能稍显暗淡，但却与公寓中的其他材料完美地融合并凸显出细节。公寓中的几个区域也是通过地板进行框定的，厨房区域的地面则用中东风格的瓷砖进行装饰。

The extensive use of materials and colors on the "white canvas" as the walls create a sense of belonging and an identity to the apartment. The parqueted Dutch wishbone flooring may be dark but brings out the rest of the material and details in combination with the apartment. Several of the areas in the apartment are framed by the flooring: in the kitchen there is rectangular frame decorated by Middle Eastern tiling while in the bathroom there are Art Deco style tiles.

黑色金属架的设计是受到一个特拉维夫街市货摊的启发。它重量轻又带有一点工业气息。开放式的设计，使厨房和客厅之间能形成有效的互动，功能分区也有了框定。

The black metal shelf, inspired by a Tel Aviv market stall, which runs between the kitchen and living area is light but also slightly industrial. Through its length, the kitchen is revealed and the work environment is framed.

照明器材和大部分家具都是在特拉维夫和国际跳蚤市场上买来的。公寓里的装饰画与住户的身份相匹配，例如包豪斯档案馆里搜集到的照片、个人的建筑素描、现代绘画以及特别受欢迎的地方的照片。很多细节装饰品和家具都有与公寓本身相仿的年头，可以追溯到特拉维夫创市之初。

Much of the furniture, such as the lighting fixtures, was collected in Tel Aviv and international flea markets. The framed images in the flat are consistent with the identity of the resident: photographs collected from the Bauhaus archives, personal sketches of buildings, modern drawings and snapshots of especially loved places. Many of the details and pieces of furniture are of a similar age to the apartment itself, dating from the early days of Tel Aviv.

# 古雅的阁楼
# Old Attic

项目地点：波兰，格莱维茨 ❋ Location: Gliwice, Poland

设计师：Dominika Trzcinska，Michal Kotlowski ❋ Designer: Dominika Trzcinska, Michal Kotlowski

摄影：Przemyslaw Skora ❋ Photographer: Przemyslaw Skora

设计公司：Superpozycja Architekci ❋ Design Studio: Superpozycja Architekci

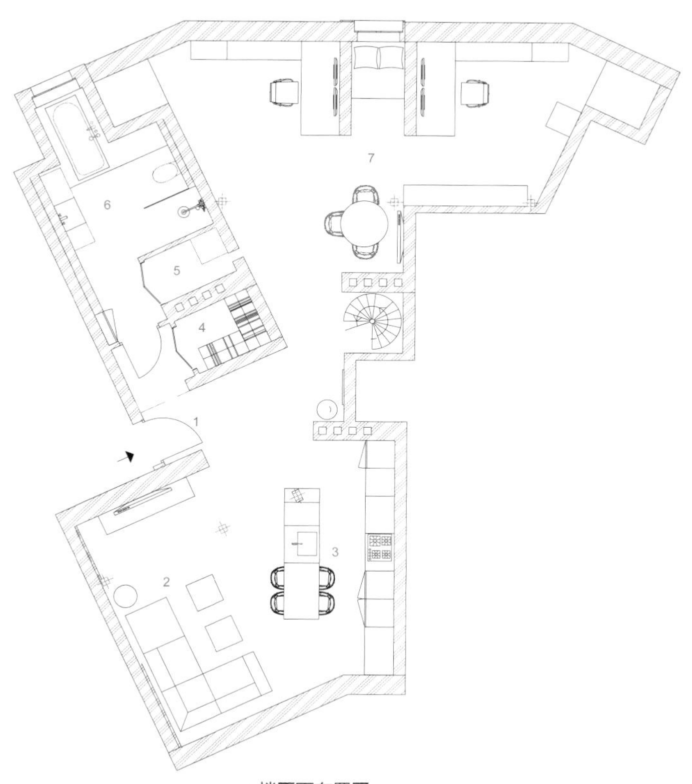


一楼平面布置图
First Ground Plan

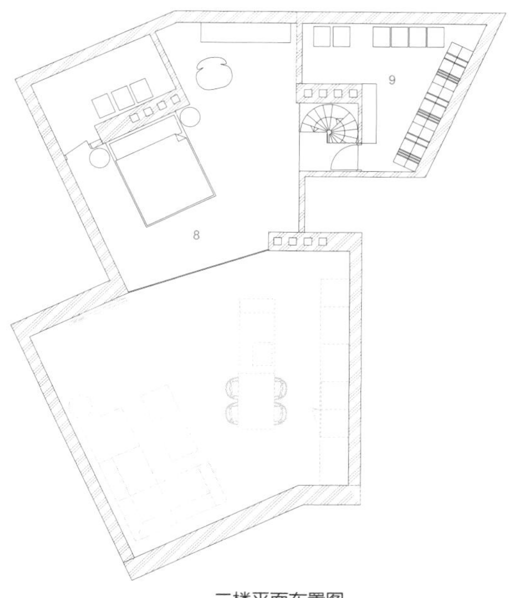


二楼平面布置图
Second Ground Plan

Superpozycja Architekci公司的两位创始人接受了这么一个挑战——将波兰格莱维茨的一个旧阁楼改造成一套崭新而又现代化的公寓。由于改造前的阁楼已破败不堪，因此这样的改造确实是一次严峻的挑战。第一步就是尽可能地使用空间来建立一个功能系统。凭借层高的优势，建筑师设计了一个多出两个房间的夹层。

Two founders of Superpozycja Architekci took up the challenge transforming an old attic in Gliwice (Poland) to a brand new, modern apartment. Due to the fact that the attic was a virtual ruin, such attic conversion was a real challenge. The first step was to create a functional system, using as much space as possible. The height of the room was an advantage and allows the architects to build a mezzanine divided into a couple of additional rooms.

GOROL

PLASTIC DESIGN
CLUB DESIGN

Have a seat

为了打造一个更宽敞、明亮的空间，设计师们使用了很多天窗引入自然光，让空间看起来更舒适。阁楼保留了原有的高度，空间更宽阔。建筑师们没有更换原有的屋顶结构和旧的砖石，反而回收了废旧的木材来制作桌子和台面。大部分的内部装饰都是由明亮而又简单的家具组成，且这些家具与背景融为一体。装饰画和照片等时尚元素则丰富了公寓的设计。夹层里有一间卧室、一个嵌入式的衣橱和一个储物空间。而且，他们设计了一面玻璃墙，这样从卧室就可以看到客厅和厨房。

In order to create more space and proper climate, they use many roof windows that lit the interior and made it look cozy and quite inviting. The attic kept its original height which made the interior look very spacious. The architects used the existing parts of roof structure and exposed old bricks. They decided to reuse the wood received from demolition to design wooden worktops and tables. The interior design is composed mostly of bright and simple furniture which blends into the background. Pictures and photographs are some of the stylish elements that enrich the apartment design. There is a bedroom, a walk-in closet and storage space in the mezzanine. They put there a glass wall too so that it is possible to observe the living room and the kitchen from the bedroom.

KATOWICE
GLIWICE

KATOWICE
GLIWICE

原先的洗衣间被改造成了浴室、杂物间和另外一个步入式衣橱。建筑师们决定也在浴室采用外露的砖块，并把它们涂成了石墨色。浴室里有独立的浴缸和淋浴间。在白与灰的围绕下，旧糙木看起来非常漂亮。

An old laundry room was transformed into the bathroom, utility room and another walk-in closet. The architects decided to expose bricks in the bathroom and paint them in graphite. There is a separate bathtub and shower cabin. Old, rough wood looks nicely surrounded by white and gray colors.

# 自然意象

# Deep In Nature

项目地点：中国，澳门 ❋ Location: Macao, China

项目面积：244 平方米 ❋ Project Size: 244 $m^2$

设计师：廖奕权 ❋ Designer: Wesley Liu

设计公司：维斯林室内建筑设计有限公司 ❋ Design Studio: PplusP Designers

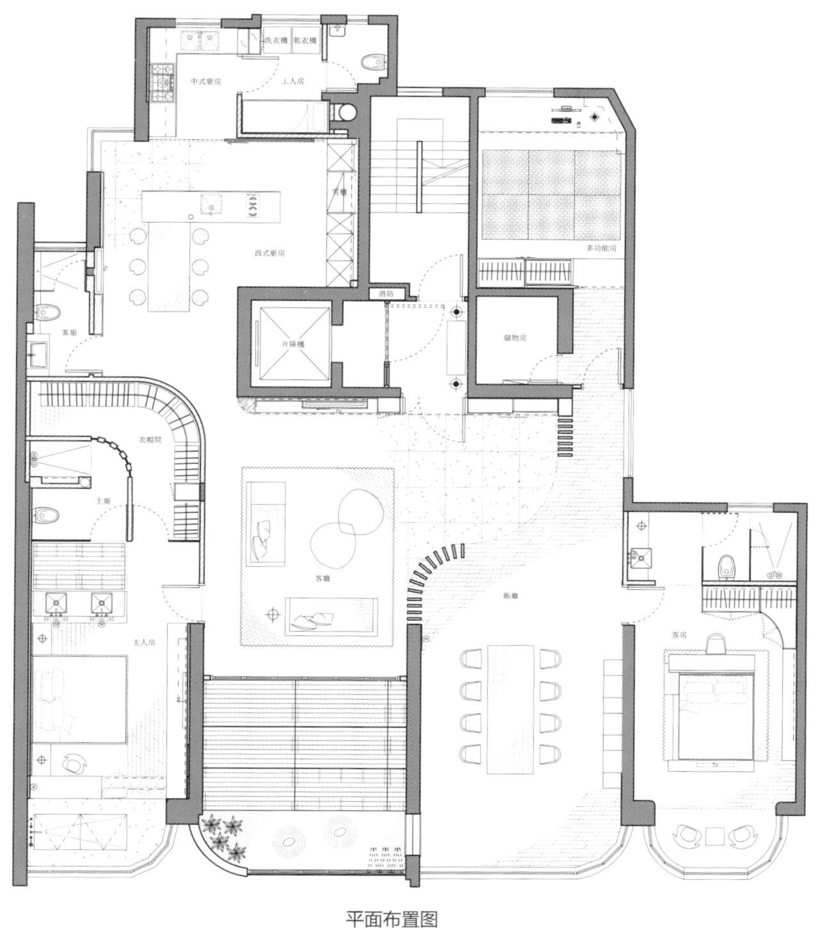


平面布置图
Apartment Plan

大自然为艺术家和设计师带来无限灵感。室内设计师 Wesley Liu 在规划住宅设计空间时，加入了自然元素，让城市里的人回家后享受回归自然的舒泰。这次，他不仅活用木材塑造树木等自然意象，更顺应结构墙的走势，在空间里融入弯曲线条，优雅动人。

Nature provides artists and designers with boundless and endless inspiration. By adding in natural elements in planning and designing of the residential space, interior designer Wesley Liu allows owners to easily enjoy the comfort and peace of nature within their own home. In this project, the designer made flexible use of wooden materials to revive various natural scenes, such as trees. Also, following the run of structural walls, he infused curved feature into the space which is therefore given an elegant touch.

自然环境与人造空间相比，前者自然气息更浓厚；室内与室外空间相比，当属室外空间更有自然氛围。因此，设计师规划此住宅空间时特意扩充露台占据的空间比例；用以标示露台范围的户外木地板，从原有露台地面延伸至客厅，令自然气息悄悄渗进室内。每当人在客厅躺卧和闲坐沙发上时，弥漫户外气氛的露台连接着室内与室外的空间，感觉轻松自在。

Apparently, a non-artificial environment bears more flavor of nature compared with a artificial environment, and the outdoor often outperforms the indoor with distinct natural vibe. Given that, the designer purposely enlarged the proportion of balcony to the space. Marking the boundary of balcony, the wooden floor outside was extended into the living room which is therefore permeated with an outdoorsy atmosphere. Lying or sitting on the sofa in the living room, residents can still enjoy the outdoors vibe from the balcony. Such special relation reveals the harmony between the indoors and outdoors, bring occupants comfort and freedom.

虽然室内并没有天然树木，但设计师巧妙运用木方，在餐厅天花上拼凑出形状不规则的立体“树枝”装置，使平淡的天花变得多“枝”多彩。设计师指出，在立体树枝之间垂下的球状吊灯象征果实，意味着树上果实累累。

Despite the absence of a real tree in the space, the designer managed ingenious use of wood and pieced together a vivid installation in the form of irregular tree branches on the dining room ceiling. Thus, the plain ceiling became vigorously "branchy". According to the designer, the ball shaped pendant lights suspending among the stereo branches are symbolic of fruits and imply a fruitful tree.

设计师在规划功能分区时，巧妙地融入了弧线。由于入口处的结构墙不能拆掉，所以客厅、餐厅难以用常规的直线方式一分为二，于是设计师在平面图上利用曲线，顺势将客厅与邻近厨房的储物室连贯。这道以弯曲走势发展的流利线条，在居室的立体空间内，衍生出弧度相同的假天花，地面则以不同深浅色调的木纹地材装饰，形成弯曲的分界线，同时在分界处以一块块顶天立地的木板疏密有致地围绕饭厅，区分出客厅、餐厅两大功能区域。此外，为呼应上述弯曲线条，分隔厨房和主人房的墙壁亦采用弧形造型。

In nature, curved, tender and organic lines seem everywhere while stiff and straight lines are relatively rare. Instead of taking deliberate references to those organic lines, the designer skillfully integrated curves in divisions planning and the establishment of boundaries. As the entrance close to an irremovable wall of structure behind, it is not likely to distinguish the living and dining rooms with straight lines. A curved line is therefore employed to connect the living room with the storage room adjacent to the kitchen. This beautiful and fluid curve brings out an artificial ceiling with the same arc in the space, and the boundary between the living and dining rooms. Grained flooring of different shades covers the two rooms, forming a curved boundary. The wooden boards were stood up well-proportionately, having the dining room partially surrounded and marking out boundary. Also, in order to correspond with this curved feature, an arced wall is used to separate the kitchen and the master room.

开放式厨房餐桌以天然实木做台面，身在其中仿佛在树下进餐。玻璃门内是密封式厨房，可以明火煮食。

The dining table in the open kitchen is covered with natural solid wood, generating a sense of dining under the tree. Behind the glass door is the closed kitchen for cooking.

主人房电视背景墙以无数实木小块拼砌而成，效果天然，散发着自然气息。木装饰旁边的黑板上写着："家不是仅仅是一个住所，而是一种家的感觉"，点出住宅设计的要旨，设计师不仅期望塑造一个空间或住所，而且是营造一种感觉或氛围。

The television wall is not distant from the end of bed in the master room. The wall covered with innumerable pieces of solid wood creates unadorned effects and natural ambiance. On the adjacent blackboard writes "Home is not a place... It's a feeling", which reflects the philosophy of this home design. The designer aims to create not just a space or place, but a feeling or atmosphere.

为提升客房收纳容量，特意升高地台作储物用途。房内地面铺榻榻米地席，营造日式茶室氛围。

The platform in the living room is purposely elevated to increase the storage volumes. Tatami mats on the floor create a vibe of Japanese tea room.

HOME
is not a
PLAC
it's a
feelin

大自然是朴实无华的，设计师特意为户主创造了质朴自然、独一无二的艺术装置，例如：位于饭厅天花的立体树枝装置，客厅主题墙——以现成材料如木材、酒塞、饭碗、书籍（包括 MH 杂志）等创造的立体装饰墙，露台墙面上以块状旧船木拼砌组合而成艺术挂饰，还有用来装饰厨房壁面的挂画等。

Inspired by the simplicity of nature, the designer deliberately created plenty of plain, natural and unique artistic installations for the owner, including the stereo installation of tree branches on the dinning room ceiling and the themed wall in the living room, which is made of established materials such as wood, wine bottle stoppers, bowls, books (inc. MH magazine). The artistic pendants suspending from the balcony wall are made with blocks of old ship wood. On the kitchen wall hangs a unique decor painting. Designed by the designer, all of these were given distinctive styles and quite efforts.

# 曼哈顿海滩别墅

# Manhattan Beach Residence

项目地点：美国，南加利福尼亚 ❋ Location: Southern California, United State

项目面积：418 平方米 ❋ Project Size: 418 $m^2$

设计公司：DISC Interiors ❋ Design Studio: DISC Interiors

本案位于南加利福尼亚迷人的曼哈顿海滩上，总面积约为 418 平方米，内含 5 个睡房和 5 个浴室。本案通过大量纹理丰富、气质纯朴的布艺和纺织材料，集中展现了当代建筑的设计精髓。设计师从众多工匠处找来了皂石、大理石、图形水泥和瓷砖，并与木工合作，精心打造了本案的所有家具和建筑细节。同时，设计师与油漆匠、陶艺师和艺术家合作，在本案各处摆放了多种风格统一的艺术单品，为本案营造了一种独特的感觉和温暖氛围。虽然本案位于美国，但是其总体设计和细节无一不流露北欧风格设计的特点。

This home is approximately 418m$^2$. It is a 5-bedroom, 5-bathroom home in Manhatton Beach, which is an incredible beach town in Southern California. The home is a merge of contemporary architecture, with textural and rustic materials. The designers sourced beautiful soapstone, marble, graphic cement and ceramic tiles from artisans, and worked with carpenters to execute all of the custom cabinetry and detailing for this home. The designers also worked with painters, a ceramicist, and artists to create one of a kind works that are installed throughout the home that give the home its unique feel and warmth. Although the house is built in the US, there is a warmth and a nod to Scandinavian design in its detailing and overall design.

本案的设计灵感源自加利福尼亚海岸线风景的颜色和脉络，并充分利用了白天不同时间反射进来的光线方位变化。曼哈顿海滩与一个美丽的码头相毗邻，是个充满生机活力的地方。但是，设计师想去掉海滩建筑里一些常见的海上元素，从而引入一些更能表达大海和海岸线颜色的元素。同时，设计师也在观赏当代的海洋照片中受到启发，特别是摄影师杉本博司(Hiroshi Sugimoto) 描绘海洋和地平线那一系列的“海景”。尽管这些作品是黑白的，但是它们充分表现了光线的渐变、色调的深度，以及它们能呈现的复杂和深度。在概念阶段，设计师就希望把黑、白色调以及两者之间的渐变色囊括到设计中。

The inspiration for this home was the California coastline, the colors and textures of the landscape, and the way the light reflects at different times of the day. Manhattan Beach is a vibrant and active beach community with an incredible pier, but the designers wanted to move away from the nautical themes that are often seen in beach homes, and more into a direction that is inspired by the colors of the sea and coastline. The designers were also inspired by looking at contemporary photography of the ocean, specifically the artist, Hiroshi Sugimoto's "Seascapes" series, which are black and white photographs of the ocean and the horizon. Although these works are black and white, they showcase the depth of tone and gradual shifts of the light, and complex and exciting that can be. Early on, the designers wanted to incorporate tones of black and white, and a neutral tonal color palette taken from both of these directions.

在设计的初期，设计师决定在一楼采用涂黑的橡木做橱柜，而浅白色的橡木则做地板。厨房和客厅相连，采用开放性设计，客厅边上是一道 6 米宽的滑门，打开后能看见别墅的后露台和泳池区。对于厨房的设计，设计师的目标是使涂黑橡木橱柜所带来的厚重感与厨房内由艺术家 Forrest Lesch Middelton 设计的瓷砖拼贴墙壁、酒吧椅的黄铜色调，以及厨房的整体照明相互融合，不显突兀。此外，设计师还选用天然皂石来做柜台，上面的涂油变黑以后，会形成很美的纹路。拼贴瓷砖、自然木纹和皂石相互结合，使整个设计充满了恬静之感，赋予了原本单调的黑色以深度。此外，本案的设计还非常聪明地将其明显的现代气息与传统元素融合在一起，实现了“新旧混搭”的设计理念，这个理念也是设计师一直在追求的，并将其贯穿于设计作品中。

In the beginning design phase, the designers decided that the first floor would have blackened oak for the cabinetry combined with the lighter white oak plank flooring. The kitchen and the living room are essentially and open floor plan, with a 6-meters large sliding door that opens to the back patio and pool area. For the kitchen design, the designers' goal was to balance the richness of the black-stained oak cabinetry with the warmth of the tiled backsplash, which is by an artist Forrest Lesch-Middelton, and the brass accents through the bar stools, and lighting. The designers used soapstone for the counters which is a natural material, when oiled almost turns to black, and has incredible veining. Combined with the tiles, natural wood grains, and the soapstone, there is a lot of "silent patterns" at work, that really give the blackness a sense of depth. Design wise, the designers really wanted to merge this sense of modernity with tradition, this idea of "old meets new." In our work, the designers are always attempting to walk this line between the two.

SEE

# 圣塔莫尼卡别墅
# Santa Monica Villa

项目地点：美国，加利福尼亚州 ❊ Location: Santa Monica, California

设计师：Krista Schrock，David John Dick ❊ Designer: Krista Schrock and David John Dick

摄影：D.Gilbert ❊ Photographer: D.Gilbert

设计公司：DISC Interiors ❊ Design Studio: DISC Interiors

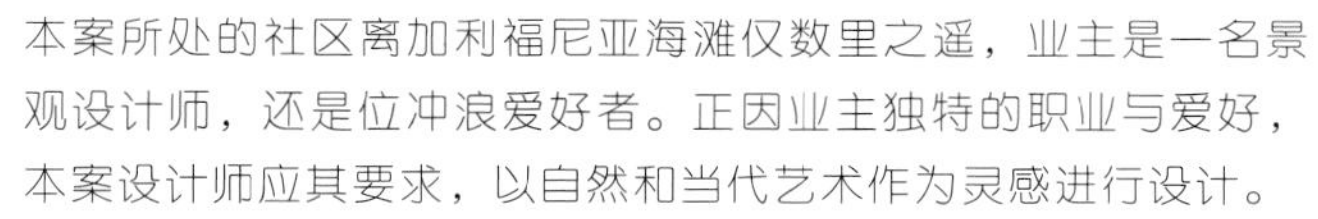

本案所处的社区离加利福尼亚海滩仅数里之遥，业主是一名景观设计师，还是位冲浪爱好者。正因业主独特的职业与爱好，本案设计师应其要求，以自然和当代艺术作为灵感进行设计。

This home is in a residential neighborhood a few miles from the California beach, and the clients requested a home that was inspired by nature and contemporary art, as they are landscape designers, and active surfers.

本案建筑由 JFAK 建筑事务所设计，室内设计师的设计目的是要凸显建筑的特色，同时为室内空间制造温馨的氛围，增加空间质感。设计师为本案的设计找来了各种艺术品和具有独特装饰的摆设，如复古的铜器、粗陶器、黑钢、缟玛瑙、假橡木装饰等，并使整个设计杂而有序。

The designers' goal for the home was to accentuate the architecture by JFAK Architects, while creating warmth and texture to the volumes of space. For the interior design of this home, the designers sourced works and furnishings that have unique finishes and add a subtle complexity to the interior spaces, such as vintage brass, rustic ceramics, blackened steel, onyx, and ebonized oak finishes.

由各种几何形构成的架子墙栉比鳞次、形状迷人，设计师为此从当地的陶器匠人手中找来了各种陶器，还有大小不一的雕塑来制造一个平衡而又不平衡的感觉，正如建筑师给这幢建筑的设定一样。

As the shelving units are beautifully over-scaled and geometrical, the designers sourced ceramics from local ceramicists and sculptural works of different scale to create a balance, and imbalance that the architects had established with the architecture.

设计师在客厅采用了模拟星空的照明，通过向多个方向旋转的照明增强空间。沙发面对花园，而不是壁炉，因此室外景观和室内并重。

The designers selected a living room lighting fixture that mimics the stars at night, and spans in multiple directions to accentuate the architectural volume. The designers also designed the custom sofa to face the gardens instead of the fireplace, so the emphasis is on the exterior landscape, as well as the interior.

本案的业主是景观设计师，她设计了一个迷人的户外花园。从家里往外一看，就是这片区域。每天下午，餐厅的光源极其丰富，针对这一特点，设计师在自然光主导下，加入了太平洋的颜色和色调——靛蓝装饰。在主人房中，设计师采用了亚麻的墙布和窗帘以增加房间的质感和深度。

The client is a landscape designer, and she designed an incredible outdoor private area that this home looks out onto. The home receives an incredible amount of light in the afternoon, so the color palette the designers selected is a response to the light. The color palette is driven by natural tones, ornamented with deep indigo blues, pulling the tones and colors of the Pacific ocean. For the master bedroom, the designers brought in grass cloth wallpaper and drapery to add texture and depth to the room.

本案设计师认为室内设计不一定要有鲜艳的颜色和抢眼的图案。家应该是让人有灵感的地方，但是也要让人感到舒服和享受。设计师想让他们的客户回家就是回到一个放松的地方，一个与家人和亲友相聚的空间。

The designers believe interiors do not have to be loud in color and pattern to be powerful. Our homes should inspire us, but they should also provide comfort and nurture. The designers want their clients to come home to their spaces to relax, and to have unique rooms to be with their family and friends.

设计师希望室内空间感觉舒适，而且能展现出主人游历广泛和喜欢收集藏品的爱好。设计师很欣赏自己在室内设计中使用的各式各样纹理，例如：亚麻布、天鹅绒，它们是天然又结实的材质，以大胆的图案凸显其特色。设计师特地为本案找来复古的地毯与现代建筑混搭，平衡了地毯粗糙的线条，并界定起居范围，纤维织物和地毯衬托出了艺术品的质感。本案含有二十世纪六十年代的现代主义元素，同时混合了加利福尼亚海滩休闲的感觉。设计师想打造一个休闲的居家环境，在白天通风、空旷，在夜里又能使居住者放松下来，尽情娱乐。

设计师为了让房间中素静的色调显得更饱满，采用了精致的材料进行涂层。设计师都倾向于极简抽象派画家，如艾格尼·马丁，罗斯科和罗伯特·雷曼，也很喜欢日式设计。设计师运用自然界可以看到的色彩，如草原或草地上常见的颜色。

The designers adore spaces that look well-traveled, and casually collected, and comfortable for our clients. The designers appreciate a variety of textures in their interior spaces, and specifically in this home. They sourced linens, velvets, natural textures, solids, and accentuated with bold patterns. The vintage rugs the designer sourced for this home mix with the contemporary architecture, balancing the hard lines, and help to define the living areas. The fabrics and the rugs also complement the art works, and the home has elements of 60's modernism, mixed with a California coastal luxurious feeling. The designer's goal was to create a casual home that felt open and airy in the daytime, but one that also transitioned to the evening if the clients are entertaining, or simply relaxing.

The designers are drawn towards rooms that are rich in neutrals, and layered with subtle textures and materials. They have always been drawn towards minimalist painters such as Agnes Martin, Rothko, and Robert Ryman, as well as Japanese design. The designers utilize a palette that is seen in nature, such as subtle shifts of color you might see in a prairie or grassland.

# 寿·森林

# Cape Mansions

项目地点：中国，香港 ※ Location: Hong Kong, China

设计师：廖奕权 ※ Designer: Wesley Liu

设计公司：维斯林室内建筑设计有限公司 ※ Design Studio: PplusP Studio

材料：清水混凝土、木头、铸铁、不锈钢 ※ Material: Bare Concrete, Wood, Cast Iron, Stainless Steel

台灯、灯笼、布艺和 ※ Lamps, Lanterns, Cloth Art and

家具设计：许伟彬，曹世妹 Furniture: Hsu Wei Bin, Tsao Shih Mei

客厅连接露台，尽揽蓝天碧海。家具、灯饰绝大部分以废弃木材、竹藤、铁枝等构思构造，突显质感纹理及线条结构，每件都是造型独特的手工艺品。客厅的水泥墙是设计师最喜爱的部分。设计师故意保留一小部分的平滑面，目的是要营造出自然与历史的痕迹感。一个天然和环保再生材料制作而成的永恒的空间。

The living hall is connected to a balcony where you enjoy a view of blue sky and sea. Most of the furniture and lights are made of obsolete wood, bamboo vines and iron wires, highlighting the texture, pattern and lines. Each of them is a handicraft with a unique form.The favorite part of the designers: the cement wall in the living room. The designers reserved a small part of smooth surface to create historical traces on purpose. Natural, environmentally-friendly and renewable materials were used to form such an eternal space.

家具的材料几乎都是由两位艺术家许伟彬和曹世妹回收来的。两位艺术家善于取材，他们收集废料如木头、铁管、不锈钢、竹和塑料等加以创作，为作品赋予功能、价值。客厅的沙发、扶手椅、茶几、影音架，饭厅的餐桌、餐椅、矮凳。部分旧物料由于依然完好，不影响整体设计效果，设计团队便沿用不改，像入门区、露台、厨房的板石，既环保，又可留住历史足迹。

Nearly all the furniture was recycled by two artists, Hsu Wei Bin and Tsao Shih Mei. Skillful in material selection, the artists gathered obsolete materials like wood, iron pipes, stainless steel, bamboo and plastic and transform them into functional and valuable artworks, including the sofa, armchair, tea table and video frame in the living room and the dining table, dining chairs and low stools in the dining room. Some obsolete materials were well preserved and did not affect the entire design effect, so the designers kept them, such as the slabstone of entrance, balcony and kitchen which are environmentally-friendly and indicate the trace of history.

日光映照下的家具陈设显得立体而富有质感，入门一侧的格子屏风是由磨砂玻璃搭配了旧木框架制成，细看能看到上面的残余油漆。考虑到使用的舒适度，长凳及扶手椅多附带座垫、靠垫。斑驳的混凝土墙历经岁月，饱含了旧日装饰所留下的痕迹。

In the sunlight, the furniture dimensional appearance shows a vibrant look. The latticed screen at the entrance is framed with obsolete wood, and a closer look would show you the residual paint on it. In the consideration of comfort in use, the bench and armchairs are equipped with seat and back cushions and furnished with diverse patterns and textiles. The mottled concrete wall shows signs of the time-honored original decoration.

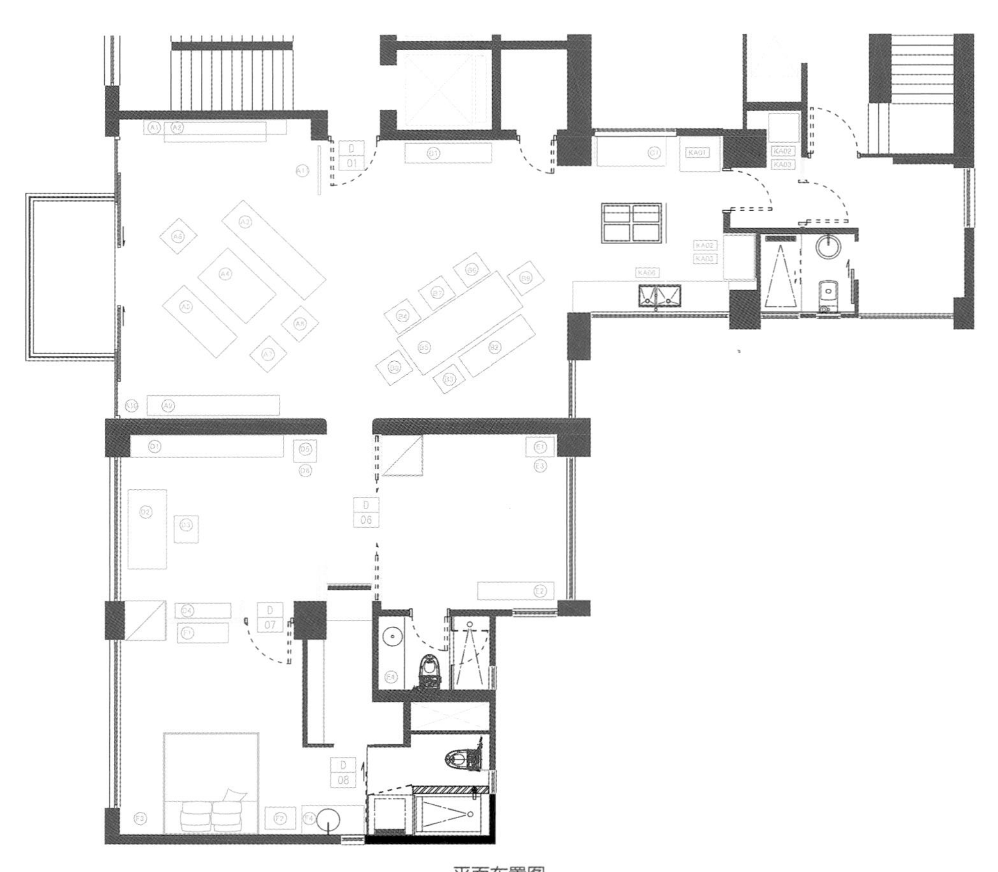

平面布置图

Apartment Plan

家具的造型极具变化，如饭厅除了两张高背椅配对成双，其余的椅凳亦各有形态，流露不同韵味。灯饰种类亦不少，饭厅吊灯便利用铁枝和铁线网弯曲成弧形，设计带着怀旧情绪，散发工业味道。此外，两厅没有装置灯槽，射灯沿方形路轨排列，构成工整的照明框架。值得一提的是户主喜欢空间开放明亮，着重灵活间隔，室内不但倚重趟门，也用了不少麻布布帘，尽量让阳光景致落入室内。毗邻饭厅的厨房，透过富弹性的木格子门作分隔，而且材质一致，所以构图才如此整齐纯净。

The furniture comes in various shapes. For instance, except the paired wing chairs, all the chairs in the dining room takes various forms and convey different charm. Moreover, there are a wide range of decorative lights. In the dining room, the pendant lamp framed with iron sticks and iron gauze is in an arched shape and looks like a nostalgic design conveying industrial charm. Additionally, no light trough is equipped in the dining room and the living room; instead, the spotlights were arranged in the lines of a rectangle forming a neat lighting framework. In particular, the owner is fond of an open and bright space and focuses on flexible partition, so sliding doors dominate the interior space, along with curtains made of linen, to introduce sunshine. The kitchen is separated from the dining room by elastic wooden latticed doors which are made of the same materials and thus contribute to a neat and clear picture.

室内空间的比例设计很灵活，公用空间以厨房为例，不但安装了折叠推拉门，而且挂设布幔，可与饭厅或融合或分开。主卧房内的衣帽间和浴室均以布帘代门，营造柔和气氛。

The partition arrangement in the interior space is flexible. In the case of the kitchen, a public space, the sliding doors and the drape can combine the kitchen with and dining space as well as separate the two from each other. The cloakroom and the bathroom in the main bedroom are equipped with drape rather than a door to create a soft atmosphere.

书房的开放层架贴着两侧墙身装设，书桌跟椅子朝窗斜放，书架与桌椅的用料、风格相似。特别有趣的是由铁线扭曲而成的时钟，带有明显的手制痕迹。主卧房处于房间区末端，跟书房只有一门之隔，左边是房间走道的艺术作品，勾起枯藤老树昏鸦的意境。短短的房间走道以白乳胶漆、水泥、木门板营造远近距离及光影明暗对比。

The open shelf in the study room is placed against two walls, and the desk and the chair face the window and share the same materials and style. What is especially interesting is the clock made of twisted iron wire, which is an obvious sign of handmade product. The main bedroom is situated at the end of the living area and is partitioned from the study by a door. On the left is an artwork in the walkway which is a reminder of a scene featuring withered vines, old trees and crows at dusk. White latex paint, cement and wooden slab door in the short walkway create the effect of near-far distance and the bright-dark comparison between light and shadow.

主卧房被深浅不一的灰调笼罩，床头墙与洗面台由水泥涂抹制而成，散发着原始纯朴的味道；木箱收集来后装上脚轮，成了特制的床头几；地灯、桌灯与吊灯由多种物料组装，别出心裁。

The main bedroom is covered with different shades of grey. The bedhead wall and the washbasin are painted and modeled with cement, conveying a sense of primitiveness and plainness. The recycled wooden box was equipped with the truckles to serve as a special night table. The floor light, the table light and the pendant are made of various materials and try to be different.

休闲室的布置质朴、自然。实木制的长木案，地面铺上动物毛皮，树干造的矮凳，焕发自然气息。

The leisure room is made highly accessible. The long desk made of solid wood, the animal fur on the floor and the cutty stools made of trunk all gives a sense of nature.

浴室延续灰色调，卧室与浴室之间是呈曲尺形的衣帽间，借着深灰色的布帘分隔。

The bathroom carries forward the grey tone. Between the bedroom and bathroom is a bevel square-shaped cloakroom which is separated the inside from outside by a dark grey curtain.

浴室同样表现水泥效果，门前是圆形的水泥制洗面盆，然后是相对宽阔的淋浴间，涂抹水泥的墙壁上贴有白底蓝花的瓷砖做装饰。

The bathroom also shows the effect of cement. Before the door stands a round cement washbasin. In the relatively broad shower enclosure, the cement wall is decorated with ceramic tiles patterned blue flowers against white background.

vivo
vives
vive
vivimos
vivís
viven
Christian Dior

# 北欧现代风格
# Modern Nordic Design

## 风格概述 Style Description

早在 20 世纪初，以芬兰、丹麦和瑞典为代表的北欧设计就开始了与欧洲同步的现代设计实践，追求传统手工艺与现代工业设计的融合，北欧现代设计是在本土的传统人文下将功能主义和现代主义设计相结合的简洁设计，从而产生了一种更富情感化的设计。就风格而言，北欧设计是功能主义的，主张实用至上。而北欧现代设计与其相比，家具的几何形式被柔化了，边角被处理成 S 形曲线或波浪线，使形式更富人情味，朴素而有机的形态及自然的色彩和质感在国际上大受欢迎。

Early in 20th, the Nordic style which represented by Finland, Denmark and Sweden, started modern design practice in order to synchronize with Europe. The pursuit of the integration of traditional handicraft and modern industrial design, Scandinavian modern design is the style which based on the traditional design and combines the functionalism and modernism in a concise style. It created a design with richer flavors.

## 细节设计 Detail Designing

北欧现代风格中整个空间的软装设计充分体现出实用、简洁的个性化特征。色彩通常以白色为基调，配合浅木色，局部使用高明度、高纯度的色彩加以点缀。空间的功能比较多，如现代居室强调视听功能或自动化设施，家具、家用电器成为主要软装陈设，室内艺术品则多为抽象艺术风格。由于现代风格室内线条简单、装饰元素少，软装需要沙发靠垫、餐桌桌布、床上用品、窗帘、绘画及艺术品等完美的搭配才能显示出美感。

The soft decoration design in the whole Nordic style space can fully reflects the practical, simple and personalized features. White is usually used as the key tone, with light wood color. Local decorated with high purity and highlight color, matched with light wood color. Modernism has relatively more functions. Media and automatic facilities, furniture, family electricity are the main soft decoration. The interior art works are mainly artistic style. As the modern interior lines are simple and with less decorating elements, soft decorating needs sofa cushions, table cloth, beds products, curtains and paintings or other art works to present an artistic elegance.

## 设计技巧 Design Techniques

▶ 色彩通常以白色为基调，最经典的色彩搭配是黑白搭配，少量的高明度的饱和色彩能够起到点睛的效果。

Generally use white as the key tone, and the most classic color is black mixed with white. A little fresh and highlighted color can add a perfect decoration to the whole space.

▶ 除了造型简洁的原木家具还可以选择金属或者新型材料的北欧家具。

Except the concise wooden furniture, can also choose Nordic furniture with metal and new materials.

▶ 选择带有几何感的灯具，例如：圆形、半圆形、圆柱体等。

Use lamps with geometrical shapes, such as round shape, half circle, cylinder etc.

▶ 室内照明的原则是采用多光源设计，在沙发旁、桌面、植物、角落等地方设置光源，让光线反射到墙壁、天花或者地面，让空间更有立体感。

The principle of interior lightings is the use of multi-light source design, place more light at the side of sofa, desktop, plants and corners to make the lights reflects on the walls, ceilings or grounds, to make the space with more dimension.

▶ 陈设与装饰应简洁，充满现代感。抽象的装饰画、几何造型的雕塑及简洁的玻璃、陶瓷、塑料、金属装饰品都极为适合。

Furnishings and decorations should be simple, and full of modern sense. Abstract decorative painting, sculpture and simple geometric modeling of glass, ceramics, plastics, and metal decorations are very suitable.

# DG

## DG

项目地点：俄罗斯，莫斯科 ❄ Location: Moscow, Russia

项目面积：14 平方米 ❄ Project Size: 14 $m^2$

设计师：Alexander Malinin, Anastasia Sheveleva ❄ Designer: Alexander Malinin, Anastasia Sheveleva

设计公司：INT2 Architecture ❄ Design Studio: INT2 Architecture

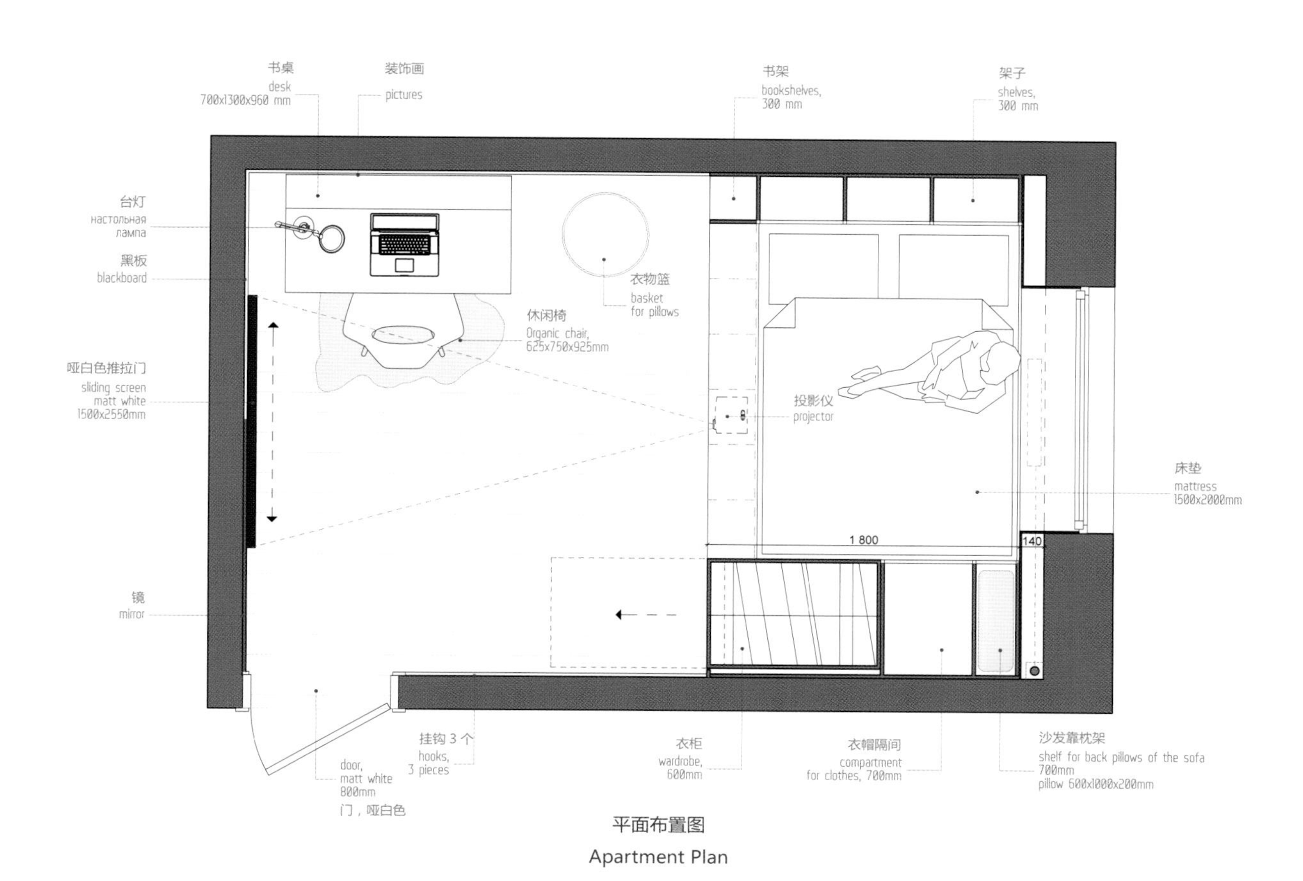


平面布置图

Apartment Plan

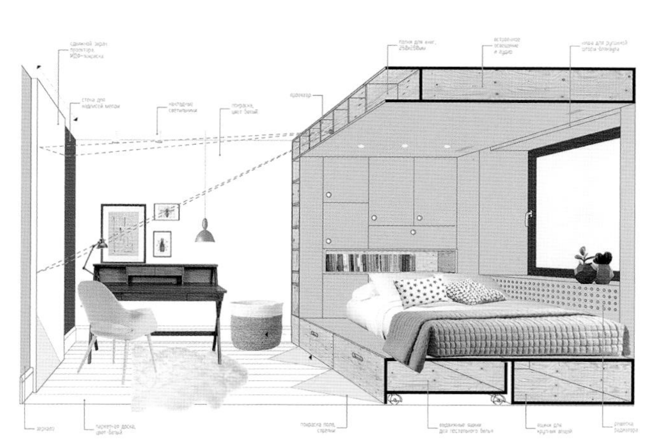

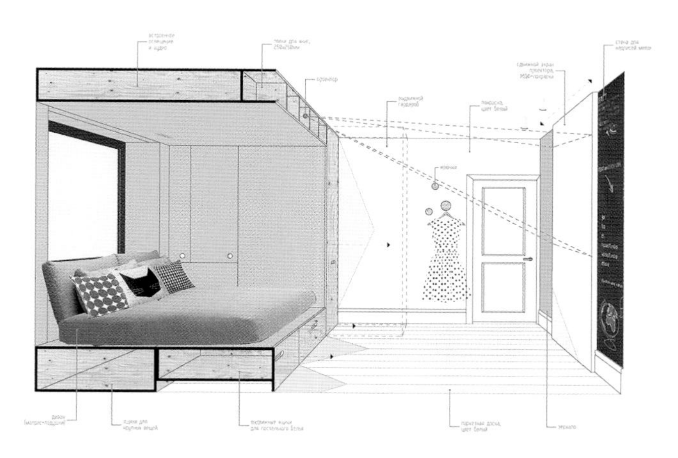

本案是专为一位 14 岁的女孩设计的小屋，她每天都生活在这个舒适而又时尚的环境中。在装修之前，房间里满是一些单独摆放的物品：存放全家人衣物以及床单枕套的衣橱、书柜、存放小物件以及书本的抽屉柜、一张单人床和一张书桌，这使整个空间没有一种整体感。设计师计划用半个房间将所有单独摆放的物件整合在一起，这样房间可以腾出一半，开辟一块新天地。为此，设计师设计了一个多功能“盒”，其包括一个紧凑型的储藏系统：衣橱、抽屉柜、书架、存放织物和大块物件的抽屉，以及一个供睡眠的床。

A small room was designed for a 14-years young girl so that she can spend her everyday life in a comfortable and stylish environment. Initially there were a lot of different freestanding objects in the room: a wardrobe for the clothes and bed linen of the entire family, bookcase, chest of drawers for small items and books, a single bed and a desk. That is why the space did not feel holistic. The designer proposed combining all the free-standing objects in one half of the room, thus freeing another half of the room and creating a new space. In order to do so a multifunctional "box" was designed. It holds a compact storage system, such as wardrobe, chest of drawers, bookshelves, drawers for linens and large items etc., and a sleeping place.

vivo
vives
vive
vivimos
vivis
viven

WELCOME
loose interpretation
"CLEAN"
como
comes
come
comemos
comeis
comen
vivo
vives
vive
vivimos
vivis
viven
hello

设计师在床对面的墙上设计了一块多功能墙面，可充当镜子和屏幕。当滑动镶板滑到一端时，可以看到一块黑板；滑到另一端时，可以看到一面镜子；位于中间时，可以当作投影仪的屏幕。

On the opposite wall a multifunctional surface was designed. It became possible by the sliding panel that in one position opens up a blackboard, in another, mirror, and in central position plays a role of a screen for the projector.

# 帕纳比住宅

# The Panamby Apartment

项目地点：巴西圣保罗 ❄ Location: São Paulo, Brazil

项目面积：170 平方米 ❄ Project Size: 170 $m^2$

设计师：Diego Revollo ❄ Architect: Diego Revollo

本案是一个约 170 平方米的公寓，位于圣保罗的高档社区帕纳比。业主是一对育有两个孩子的年轻夫妇。一家四口的生活方式非常现代化。公寓原本有一个大阳台从整个客厅向外延伸，但是，考虑到圣保罗的天气阴晴不定，设计师建议用玻璃封闭阳台并在其中采用相同的装饰，从而把阳台变成客厅的一部分。封闭阳台后，一扇大玻璃窗随即形成，使光线尽情倾洒。

This apartment with 170 $m^2$ is located in Panamby, a premium neighborhood in São Paulo. It belongs to a young couple with 2 kids that have a modern life style. In the original project, there was a big balcony in the entire living room extension. But since São Paulo weather is very instable, the designer suggested that this space became part of the living, closing the balcony with glass and bringing the same architectural finishes in the entire space. With the balcony closed, a big window was created, flooding the space with light.

从采光效果与屋主的个性出发，地面采用象牙洞石地板装饰，明亮的窗户与素色地板旨在打造一个明亮的空间。天花与墙壁没有任何多余的装饰。为了使整个结构不显得过于冰冷，所有的细木工作均采用天然木材和美国橡木以浅蜜色制成。

Starting from this light and the personality of the owners, a clear floor was chosen, the navona travertine marble crude, which has a modern finish showing the irregularities of the stone that was fixed in big stones. The light of the window and the floor was thought a clean structure. And so that the structure would not be cold, all joinery was taken in natural wood, American oak, light colored in shades of honey.

得益于一个明亮、中性色的空间，设计师能够注入更具有冲击力的装饰元素。电视区成为装饰布局的出发点，为了与极简的客厅的电视背景墙相协调，局部搭配了亮丽的色彩。考虑到空间的特点，屋主的个性以及椅子的设计，亮黄色成为绝佳选择。其他的室内陈设则采用了深浅不同的灰色与米色，并搭配木质材料。

With the clear and neutral structure, it was possible to bring more impactful elements in the decor. The starting point of the decoration was the area of television. For the harmonization of television in the living room, design chairs were chosen with strong color. The color "bright yellow" was therefore a consequence of the space, the personality of the clients and also the design of the chair.

公寓的其他房间也体现出同样的现代设计理念。厨房的地板与墙面用绿色马赛克铺贴，形成一个搭配木质橱柜的“绿盒子”。厨房环境变得非常实用，并体现出居住者的生活方式。

The other rooms of the apartment reflect the same modern concept. For the kitchen it was used a tablet green glass forming a "green box" with wood. The environment became practical and reflects the lifestyle of the residents.

QUINTESSENTIALLY PURE 2010
PANTONE®
672 C
BOOK FESTAS DECORADORES
BOOK FESTAS FORNECEDORES

# 360° 住宅

# 360° Apartment

项目地点：巴西，圣保罗 ❄ Location: São Paulo, Brazil

项目面积：130 平方米 ❄ Project Size: 130 $m^2$

设计师：Diego Revollo ❄ Designer: Diego Revollo

本案位于巴西圣保罗的 Alto da Lapa，面积 130 平方米。业主是一位与父母居住在一起的单身律师。设计师通过与业主的交流，发现了本案原有布局上的一些缺点，例如，厨房地面的设计和位于阳台另外一侧的浴室，并重新进行了改造。

This apartment of 130m2 is located in Alto da Lapa in São Paulo, it belongs to a lawyer who purchased the property in the floor plan while living with his parents and did not think it would be married when move.The first ideas exchanged with Diego Revollo the architect hired to finalize the apartment led the client to realize the error of the previous decisions.

CENTURY

设计师认为基础的装修元素很重要，因为这些不会经常去改变，例如瓷砖。要保证环境的整体性就要用更有特色和价值的瓷砖来装饰，独具特色的装饰材料让整个空间更具有整体性，也为本案中性的装饰基调奠定了基础。

Diego Revollo explained the customer to keep the integrated environments need a more noble porcelain. The material integrated spaces and defined a more neutral decor. In contemporary designs the basic elements are very important, and these do not change all the time. They are perennial.

NO SMOKING

Patricia Urquiola
AVEDON
bauhaus

Patricia Urquiola
AVEDON

Aesop

# 卡拉泰布里安扎私人住宅

# Private Apartment in Carate Brianza

项目地点：意大利，卡拉泰布里安扎 ❄ Location: Carate Brianza, Italy

设计师：Galleria del Vento-Carlo e Alessandro Colciago ❄ Designer: Galleria del Vento-Carlo e Alessandro Colciago

摄影：Alessandro Colciago ❄ Photographer: Alessandro Colciago

油漆匠：Paolo Proserpio 和 Niagara Detroit ❄ Printer: Paolo Proserpio, Niagara Detroit

本案位于意大利卡拉泰布里安扎，它将复古家具与定制家具完美地结合在一起。为了创造一个更为通风的环境，设计师们进行了两个微小的结构性调整，用一堵墙将厨房与客厅区分开来。另外，对浴室的布局进行了改动，设计了一个带有定制衣橱和鞋架的小走廊。客厅里摆放着一张标志性的沙发，是 Carlo Colciago 2013 年所设计的“curva”。一个一战时的旅行箱作为电视柜。音响柜也是按照客户的需求定制而成。

This small apartment in Brianza is a good mix of vintage and made-to-measure furniture. The designers set up two small structural interventions. In order to create a more airy environment, the designer built off the wall that divided the kitchen from the living area and changed the bathroom layout to create a small hallway containing a made-to-measure wardrobe and a shoe rack. In the living area find places an iconic sofa from Galleria del Vento collection designed by Carlo Colciago in 2013. The TV cabinet is a World War I trunk. Also the stereo cabinet is made to measure in order to fill customer needs.

厨房是设计师为业主量身定做的，材料包括白色假漆中纤板、钢材以及橄榄木。桌子由 Moorman 制造，而椅子则来自于品牌 Hay。墙灯为 FIAT 工厂所使用的复古式工业桌灯。

The kitchen is made-to-measure by us mixing white laquer MDF, steel and olive wood. The table is by Moorman while the chairs are from Hay. The wall lights are vintage industrial table light used in FIAT factory.

# 都灵梦幻之居

# Via delle Orfane Torino

项目地点：意大利，都灵 ❊ Location: Turin, Italy

设计师：Elena Belforte，Giusi Rivoira ❊ Designer: Elena Belforte and Giusi Rivoira

摄影：Livio Marrese ❊ Photographer: Livio Marrese

设计公司：Con3studio ❊ Design Studio: Con3studio

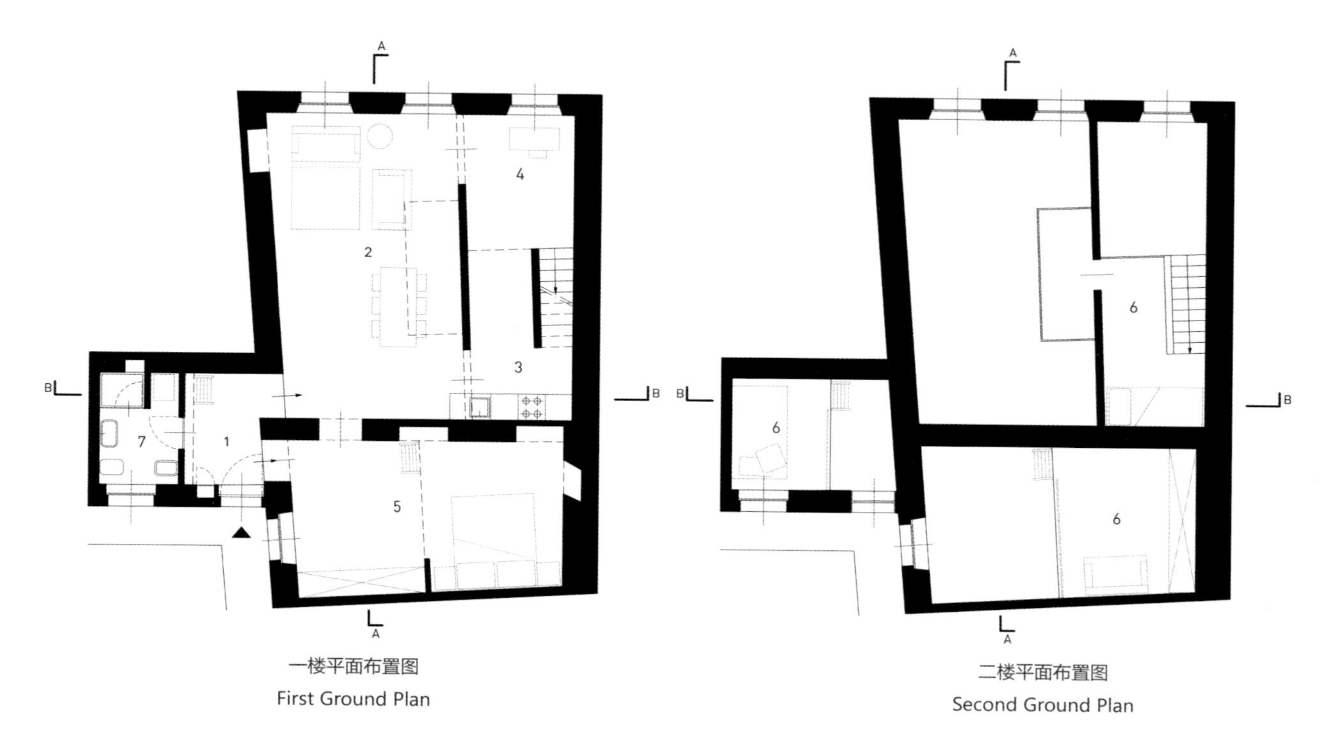


一楼平面布置图
First Ground Plan

二楼平面布置图
Second Ground Plan

本案位于意大利都灵的奥夫纳街（Via delle Orfane），参观它就像是进行一次美妙的感官之旅。穿过都灵一座历史建筑里小而明亮的庭院，跨过门槛，映入眼帘的便是一个静谧的环境，简朴却充满魅力。

First of all the apartment in Via delle Orfane is an sensory experience. Entering through a small but bright courtyard of an historical Turin building, just past the threshold, leads to a somewhat monastic setting, simple but full of charm.

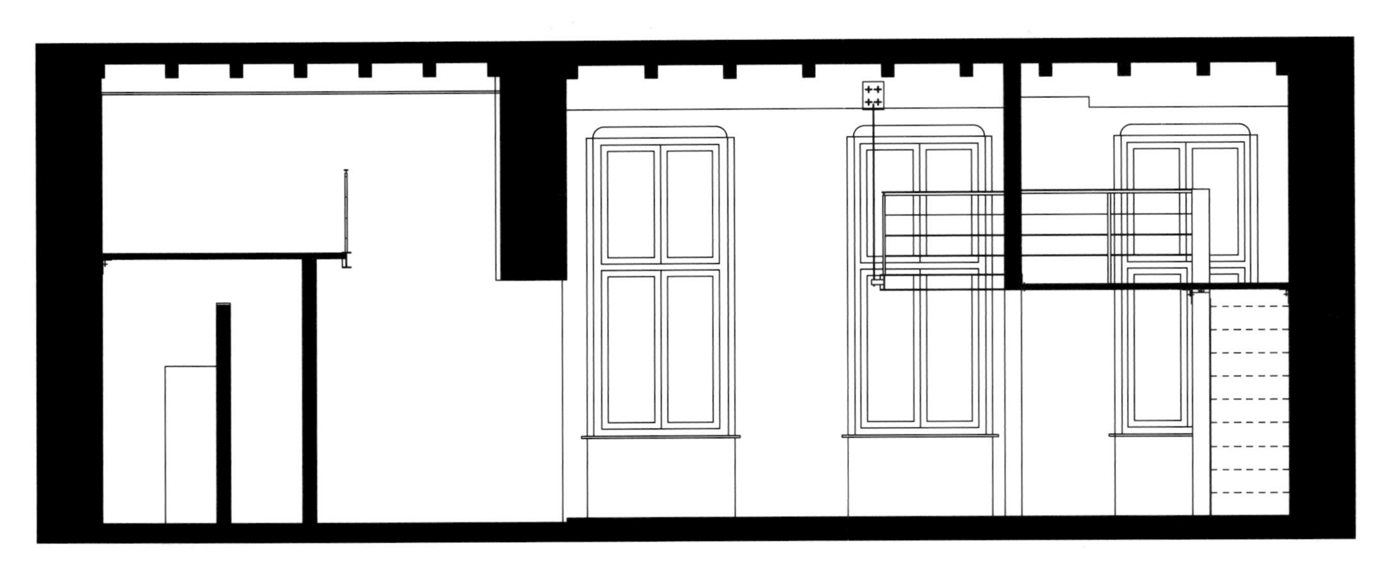

B-B 剖面图

B-B Section

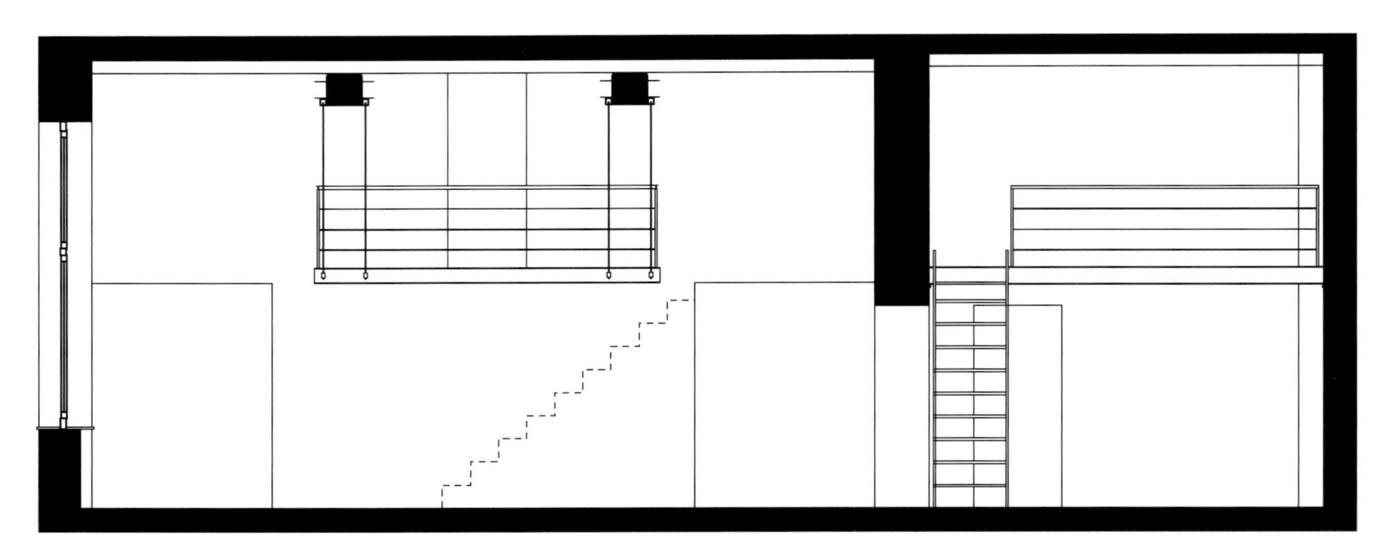

A-A 剖面图

A-A Section

夹层楼面打破了惯有的空间规律，创造出了一个新潮有趣、让人意想不到的空间，110 平方米的开敞空间让任何时候的任何活动都变得轻松自如，哪怕只是片刻的消遣时光。这里就像是可以眺望客厅的阳台，没有特定用途，却有意外之美的乐趣。本案选用的家具旨在营造特殊时空感：橱柜由保留历史气息的门改造而成；克洛普设计公司的 Cubic 咖啡桌既能变成临时的坐椅，还能重拼变成一个完整的棋盘；这里还有他们公司的 Xtable，它是一张圆角方桌，放置在 50 年代的复古餐具柜旁。古老韵味与独特设计之间呈现出强烈对比，更给人一种置身于从未见过的非凡之地的感受。

The mezzanines break the spaces regularity by creating new playful and unexpected volumes. It's 110 m$^2$ makes place for every activity, for each time of day, even for wasting time, like the balcony that overlooks the living room that has no particular use if not that of the pleasure of its unexpected beauty. The furniture has been chosen to create special moments, starting from the armoir created from salvaged historical doors and Cropdesign's coffee table Cubic that can be transformed in extra seating and can then be reorganized in a perfect chessboard. Or Cropdesign's Xtable, a square table with rounded edges, placed near a 50's vintage credenza. There are strong contrasts between salvaged and unique design, which increases the sensation of finding oneself in a place different from others, never seen before.

虽然这座公寓已经历过许多改变，但人们依旧耐心保存和修复了一些历史元素，包括十七世纪的拱形走廊与木制方格天花板。其余的改造则呈现当代风格。白色为空间的主色调，被用于所有的材料，例如石膏、木材以及铁器。

The historical elements, the few remaining after many changes that occurred over time, were preserved and recovered with dedication and patience, including the seventeenth century arched passages and the coffered wooden ceiling. The rest of the intervention is all contemporary. White dominates on everything, and is used on all materials, plaster, wood, iron.

# 米兰私人住宅

# Private Apartment in Milano

项目地点：意大利，米兰 ❋ Location: Milano, Italy

设计师：Galleria del Vento-Carlo e Alessandro Colciago ❋ Designer: Galleria del Vento - Carlo e Alessandro Colciago

摄影：Alessandro Colciago ❋ Photographer: Alessandro Colciago

灵感：附带冷色调色板的白盒子 ❋ Inspiration: a white box with a cool color palette.

本案内部设计以白色简约的线条为主。设计师将居室分为两个功能区：以实木桌为主体的用餐区和摆放着组合沙发、电视柜的客厅。电视柜是一张 5 米长的浮动床，在其下方有一个生物质壁炉和一个易于搬动的立体声组合音箱。沙发是 Designers Guild 牌的，由 100% 靛蓝色亚麻布定做而成。

A white and minimalistic line rules the interior design of this apartment. The designer split the livingroom in two functional areas: the dining area dominated by a large solid wood table and the living area, where find place the sectional sofa and the TV cabinet. Under the 5-meters. floating bed that works as TV cabinet find place a bio-fireplace and a stereo integrated solution that can be moved simply. The sofa is from Galleria del Vento collection upholstered in 100% linen from designer Guild.

所有的家具的材质均为漆白的橡木，所有门都由白漆中纤板制成，细部用以生物热橡木与锈铁处理。桌子的材质为坚实的热感橡木，而纯白色的椅子来自于品牌 Hay - About a Chair。

All the furniture are in white stained oak with doors in MDF white lacquered and details in bio-thermic oak and rusty iron. The table is in solid thermic oak while the chair, from Hay - About a Chair, complete the white on white palette.

# 我爱灰色

# I Love Gray

项目地点：意大利，罗马 ❊ Location: Rome, Italy

项目面积：80 平方米 ❊ Project Size: 80 $m^2$

设计师：Stella Passerini 和 Giulia Peruzzi ❊ Designer: Stella Passerini, Giulia Peruzzi

设计公司：Spazio 14 10 team ❊ Design Studio: Spazio 14 10 team

本案是一套重新装修的罗马公寓，“灰色”和“设计”是本案的关键词。业主喜欢艺术与设计，所以希望将自己的家变成一个更现代化和更具吸引力的空间，既可以招待朋友，又可以放松身心。还要求改造卧室，同时提高厨房的舒适度。首先，设计师移除了卧室和厨房之间的隔墙，这样既可以扩大厨房，又可以利用空间。用餐区和玄关采用了石膏天花板，以此区分开放空间的不同功能区。卧室安装的是漫射光灯，而餐桌和玄关的书架则采用了聚光灯。客户一直强调沙发的舒适性，所以设计师选择了 Ditre Italia 设计的灰布转角沙发“Kris”。除了柔和的线条以外，该沙发还在细节上体现了复古的风格。

Grey and Design are the keywords for the renovation of this apartment in Rome. The client, passionate about art and design, wanted a more modern and attractive space for her home, a place to meet friends and relax. She requested a transformation of her living room and to make the kitchen more comfortable. First, the wall that divides the living room from the kitchen was moved to enlarge the kitchen and take advantage of the space. Over the dining area and the entry a plaster ceiling was made which identifies the different functional areas of the open space, with a band of diffused light in the living room and spotlights on the dining table and along the bookshelf in the entry. A large and comfortable sofa has always been the starting point of the client, so the corner sofa "Kris" by Ditre Italia was chosen in gray linen, featuring soft lines and retro details.

本案的成功之处还在于选对了灯：Bruno Munari 在 1964 年设计的 Falkland 临时吊灯照亮了楼梯井并营造了一种令人放松的氛围；Seletti 的霓虹字母旁则是富有创意的图形；房间的正中央立着一盏抛物灯，是典型的意大利式设计；用餐区和玄关均安装了隐蔽的聚光灯，光线聚焦在漂亮的藏书上；阅读区的特色是一盏宜家的落地灯，其设计灵感来源于舞台灯光，给房间带来工业化的味道。灰色的格调、黄色光线的对比以及灯光的设计赋予了该公寓现代都市的韵味。

The set is completed by a nice selection of lights: the ephemeral pendant lamp Falkland, designed by Bruno Munari in 1964, illuminates the stairwell and creates a relaxing atmosphere; beside the neon letters of Seletti are part of a creative graphic composition; at the center of the room there is a suggestive arc lamp, an icon of Italian design; the dining area and entrance are illuminated by recessed spotlights that highlight the great library; and the reading area is characterized by a floor lamp made by IKEA, which is inspired by stage lighting and adds an industrial touch. The tones of gray, the contrast of the yellow light, and the lighting design give an urban and contemporary mood to the apartment.

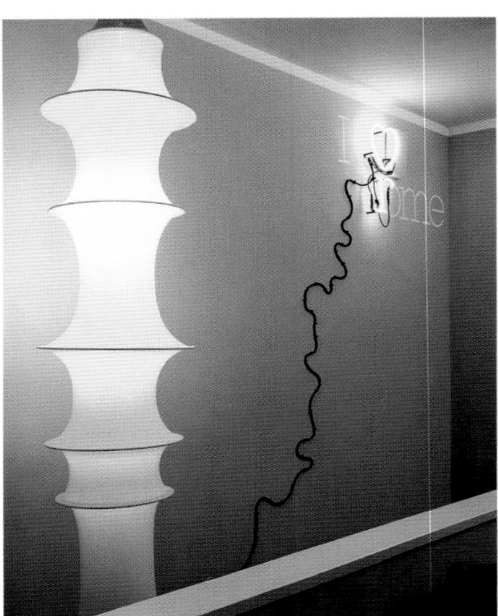

用餐区摆放着一张木制方桌，白色的玻璃将光线反射到屋内的各个角落。桌子的明朗线条让二十世纪五十年代的椅子更加显眼。在浅灰色墙面的映衬下，桌子旁边的储物柜给房子平添了复古的气息。

In the dining area is a large square table in wood and white glass that reflects light throughout the room. The clean lines of the table highlight the 1950s era chairs. Next to the table is the family kneading trough, framed by the light gray of the walls, that adds a retro charm.

立面图
Elevation Plan

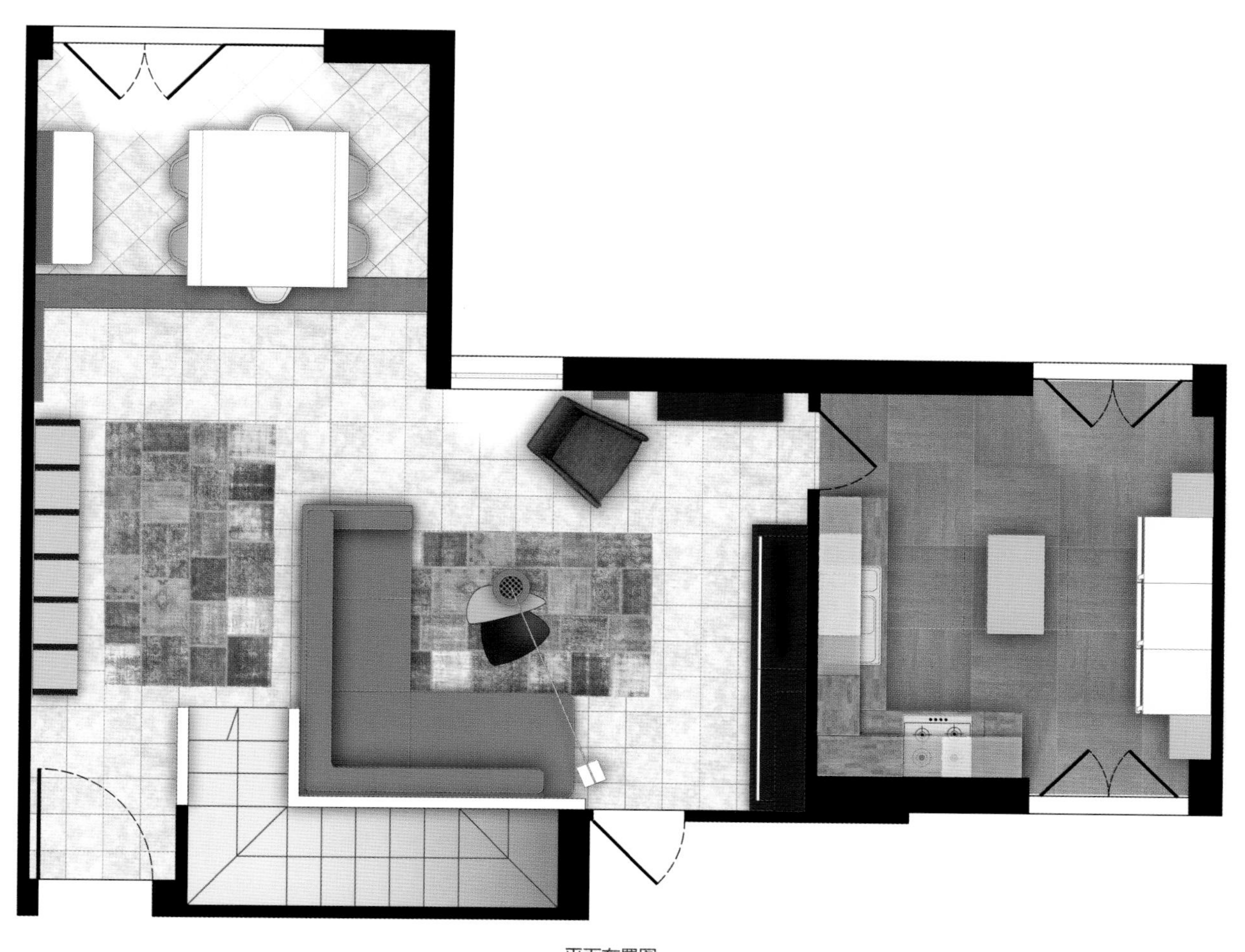

平面布置图
Apartment Plan

立面图
Elevation Plan

# NI

## NI

项目地点：俄罗斯，莫斯科 ❄ Location: Moscow Russia

项目面积：45 平方米 ❄ Project Size: 45 $m^2$

设计师：Alexander Malinin, Sheveleva Anastasia ❄ Designer: Alexander Malinin, Sheveleva Anastasia

设计室：INT2 Architecture ❄ Design Studio: INT2 Architecture

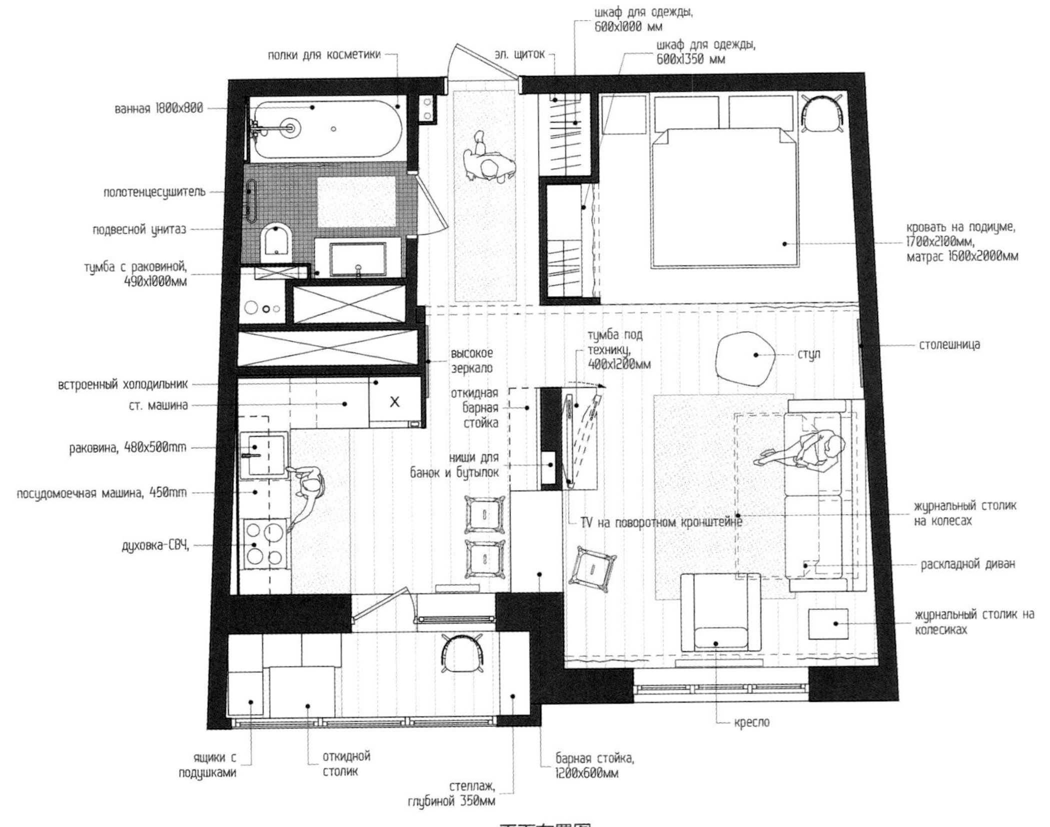


平面布置图

Apartment Plan

本案是莫斯科郊外新住房开发区中的一间单卧室公寓，面积是45 平方米。考虑到本案的面积不是很大，所以设计师的主要目的是给年轻的单身女性打造一个美观而又实用的室内环境。为此，设计师对现有的建筑平面图重新做了规划，除了浴室之外，所有非承重的隔板都被拆除，这样一来，整个公寓看起来就是一个开敞的空间，从功能上则分为客厅、卧室和厨房。

This is a bedroom apartment (45 $m^2$) in a new housing development in outskirts of Moscow. The main objective was to create a beautiful and functional interior for a young single woman taking into account the small size of the apartment. In order to do so the existing floor plan was fully reconsidered: all non-load bearing partitions except the bathroom are removed. As a result, the whole apartment works visually as a common space functionally divided into a living room, a sleeping area and a kitchen.

• Bread
• Hummus?
• Juice (tomato, apple)
I HATE STUPID
TOMATOE JUICE
МАСЛО
Don't forget to
buy more of
THESE →
BAR
КРУПЫ
OPENS here

BE full of JOY
КНИГИ
Grocery list
please buy more candy!!!
КРУПЫ
МАСЛО
Don't forget to
buy more of
THESE →
BAR
BAR OPENS here
Wine
Bottled
POETRY

空间狭小所带来的问题就是储物空间的不足，因此，设计师将该项目中唯一的一面新建的隔墙加厚，然后在隔墙的两面都设置了内嵌式衣橱。此外，设计师将床抬高，下面设置抽屉，这样就可以增加更多的空间存放大物件，同时从视觉上将卧室与客厅区分开来。最后，阳台处的座位由军用箱组合而成，这些箱子也可用于存放物件。

One of the main problems in small spaces is the lack of storage places, that is why in this project the only new partition is thickened to become two-sided built-in wardrobe. Moreover, the bed is lifted on the podium in order to get more storage space for bulky items and to visually separate bedroom from the living room. And lastly, seating places on the balcony are made of army boxes that can also be used for storage.

believe

believe

整个室内环境的颜色淡雅柔和。作为背景的白墙烘托出了其他色彩：卧室的珊瑚色、厨房的翡翠绿以及阳台的天蓝色。

he entire interior is designed in light soft colors. White walls work as a background for accent colors: coral in the bedroom, emerald in the kitchen, light blue on the balcony.

G
R
W
КАНТОВАТЬ

# MM 住宅

# Apartment MM

项目地点：德国，柏林 ※ Location: Berlin, Germany

项目面积：140 平方米 ※ Project Size: 140 $m^2$

设计师：Ester Bruzkus，Patrick Batek，Holger Duwe ※ Designer: Ester Bruzkus, Patrick Batek, Holger Duwe

摄影：Jens Bösenberg ※ Photographer: Jens Bösenberg

设计公司：Bruzkus Batek Architekten ※ Design Studio: Bruzkus Batek Architekten

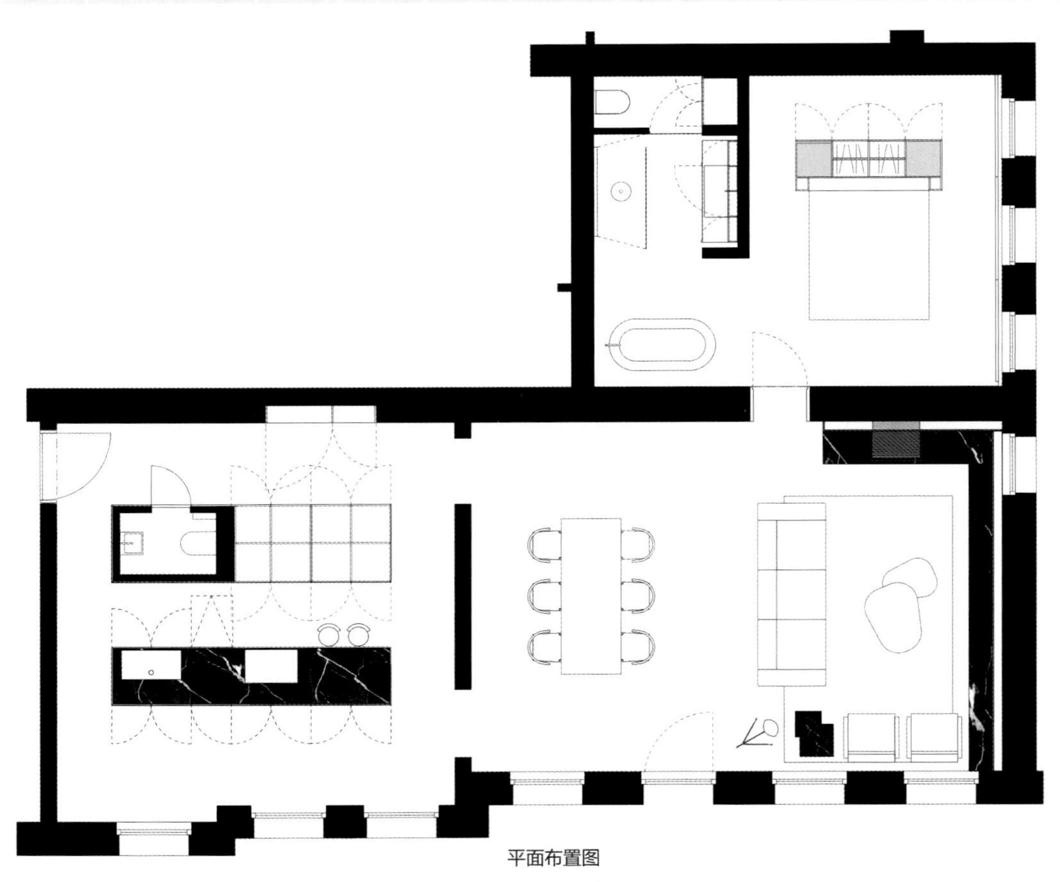

平面布置图

Apartment Plan

本案位于柏林格鲁内瓦尔德区一幢经典风格别墅的首层，Bruzkus Batek Architects 的建筑师通过剥除粉饰灰泥、雕带、镶边以及天花板的镶板等所有饰面元素，去繁就简。这么做的另外一个原因是因为这些饰面材料要么已经损坏，要么质量堪忧。

This home is located on the ground floor of a classical villa in the Grunewald district of Berlin. As part of the new design Bruzkus Batek Architects first pared the spaces down to the essence by stripping out all decorative surface elements such as stucco, friezes, trim and coffer —also since these were damaged or in pretty bad condition.

设计后的成果为格调鲜明的画廊式房间，来自 PSLAB 的简约照明理念更是将这种特点凸显出来。家具和陈设犹如展览品一般陈列在雪白的墙壁前。灰白色的水泥地搭配着岛式厨房和浴室里的光滑的白蜡木和 Nero Marquina 牌大理石，同时也映衬着客厅的长条凳。

Clear gallery-style rooms are the result, whose character is accentuated by a minimalistic lighting concept from PSLAB. The furniture and fittings are positioned like exhibits in front of the sheer white walls. A pale gray concrete floor is combined with soaped ash and Nero Marquina marble in the kitchen island and the bathroom, as well as for the long bench in the living room.

本案最主要的空间元素当属位置居中的衔接式厨房操作台，其整合式座位还可用作早餐台等。除了将所有的厨房用具藏于视野之外，设计师们还采用了无门把手设计，使设计效果更显简洁。在一体式操作系统的帮助下，轻轻一按即可打开冰箱门。一张大餐桌和一个大理石饰面壁炉与分隔墙一起把厨房与用餐区和客厅分隔开来。

The most dominant spatial element is the central and communicative kitchen block with integrated seating, for example, for breakfast on the go. The clarity of the design is underlined by hiding all ironmongery from sight and by avoiding the use of door handles. Even the fridge door opens simply via gentle pressure-and with the help of an integrated motor. A separating wall divides the kitchen from the dining and living zone with its large dining table and marble-clad fireplace.

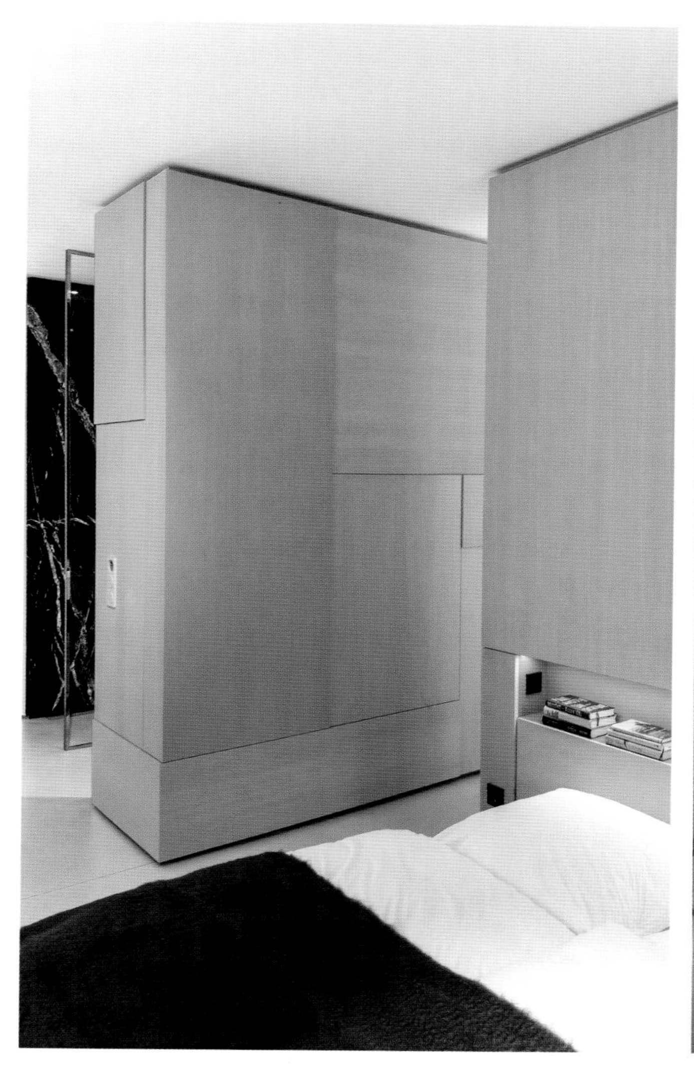

卧室与浴室相连，浴室设有直立式的淋浴间，由等高的铬合金框玻璃墙围成。光滑的铬合金架支撑着定制的韩式盥洗池，而墙壁与内嵌式壁橱则覆盖在白蜡木或大块的制式大理石板之下。

Sleeping and bathroom zones flow one into the other the bath is freestanding and the level-access shower is simply bounded by a chrome-framed glass wall. The custom-built Korean washbasin is mounted on a glossy chrome stand. The walls and built-in cupboards are clad in ash or large format marble slabs.

# 佩纳私人住宅

# Private Apartment in Paina

项目地点：意大利，佩纳 ❄ Location: Paina, Italy

设计师：Galleria del Vento - Carlo e Alessandro Colciago ❄ Designer: Galleria del Vento - Carlo e Alessandro Colciago

摄影：Alessandro Colciago ❄ Photography: Alessandro Colciago

灵感：以天然的色调营造一种舒适的氛围 ❄ Inspiration: Natural tones to create a cosy atmosphere

本案的布局很传统，空间划分很清晰。客厅以两件标志性的收藏品为主体。“Leggero”沙发是一张木架组合沙发，木架材质为着色橡木，旁边是“registro”电视柜。以实木和铁定制的桌子为厨房增添了特色。二楼卧室有一个以着色橡木为材质的环边衣橱，另一端则为一个整合式书房。定制的床也是以着色橡木为材料，并附带一个小箱子。

This apartment has a traditional layout with a clear space partition. The living area is dominated by two iconic pieces from Galleria del Vento collection. The "Leggero" sofa is a sectional sofa with a wooden frame in solid oak stained and the "registro" TV cabinet. The kitchen is enhanced by the table made to measure by designers in solid wood and iron. The bedroom on the second floor has a perimetral wardrobe in oak stained and ends up with a library integrated. The bed, designed by designers on commission, is in stained oak with a small chest.

# 闹市静所

# Inner City Calm

项目地点：澳大利亚，维多利亚省，巴拉腊特市 ❋ Location: Ballarat, Victoria, Australia

摄影：Rebecca Croft, Narelle Mann, Craig Holloway ❋ Photography: Rebecca Croft, Narelle Mann, Craig Holloway

设计公司：White Elk Interiors ❋ Design Studio: White Elk Interiors

本案位于澳大利亚巴拉腊特市市中心的交通枢纽地带，业主希望设计师把本案改造成一个市中心的宁静之所，让这个充满现代气息的空间能更有个性而又不失温暖。本案的装修过程共分为三个阶段：2012 年进行前院、后院和休闲区的建造；2013 年对主人房进行重新装修；2015 年对主浴室进行改造，设计花园与起居室。前业主在屋后建造了一个绚丽多姿的花园，这是个宁静的天堂。花园里种植了多种澳大利亚的本土花草，这也是本案设计的灵感来源。

An inner city sanctuary was requested for this home renovation and decorating project located adjacent a busy intersection in Ballarat, Australia. The designers were asked to work within the existing contemporary space to add character, warmth and impact. The project was completed in three stages and included: landscaping of the front yard and construction of a year-round living/entertaining space in the back yard (2012); redecoration of the master bedroom (2013); main bathroom renovation, nursery and living room styling (2015). Previous owners had cultivated a magnificent garden in the rear of the home, creating a tranquil haven. This garden - a mixture of Australian native flora – was inspiration for the design and styling.

皮草、奢侈陈设、环境照明与木材在本案贯穿使用，营造了温馨和主人好客的氛围。主要应用的特色材料包括柚木、循环再造金属、马赛克瓷片、粗图案墙纸、玻璃和藤垂饰照明设备。鲜花和盆栽遍布每一个空间里，它们有聚焦的作用，用有限的色彩和硬朗的外表为本案注入视觉上的情趣、柔和与生机。

Animal hides, luxe furnishings, ambient lighting and timber were used throughout the project to create cosy, inviting spaces. Key features and materials used included teak, recycled metals, mosaic tiles, bold patterned wallpaper, glass and rattan pendant lighting.In the final phase of styling, fresh flowers and potted plants were incorporated into every space. They worked as focal points, adding visual interest, softness and life to areas with limited colour or hard surfaces.

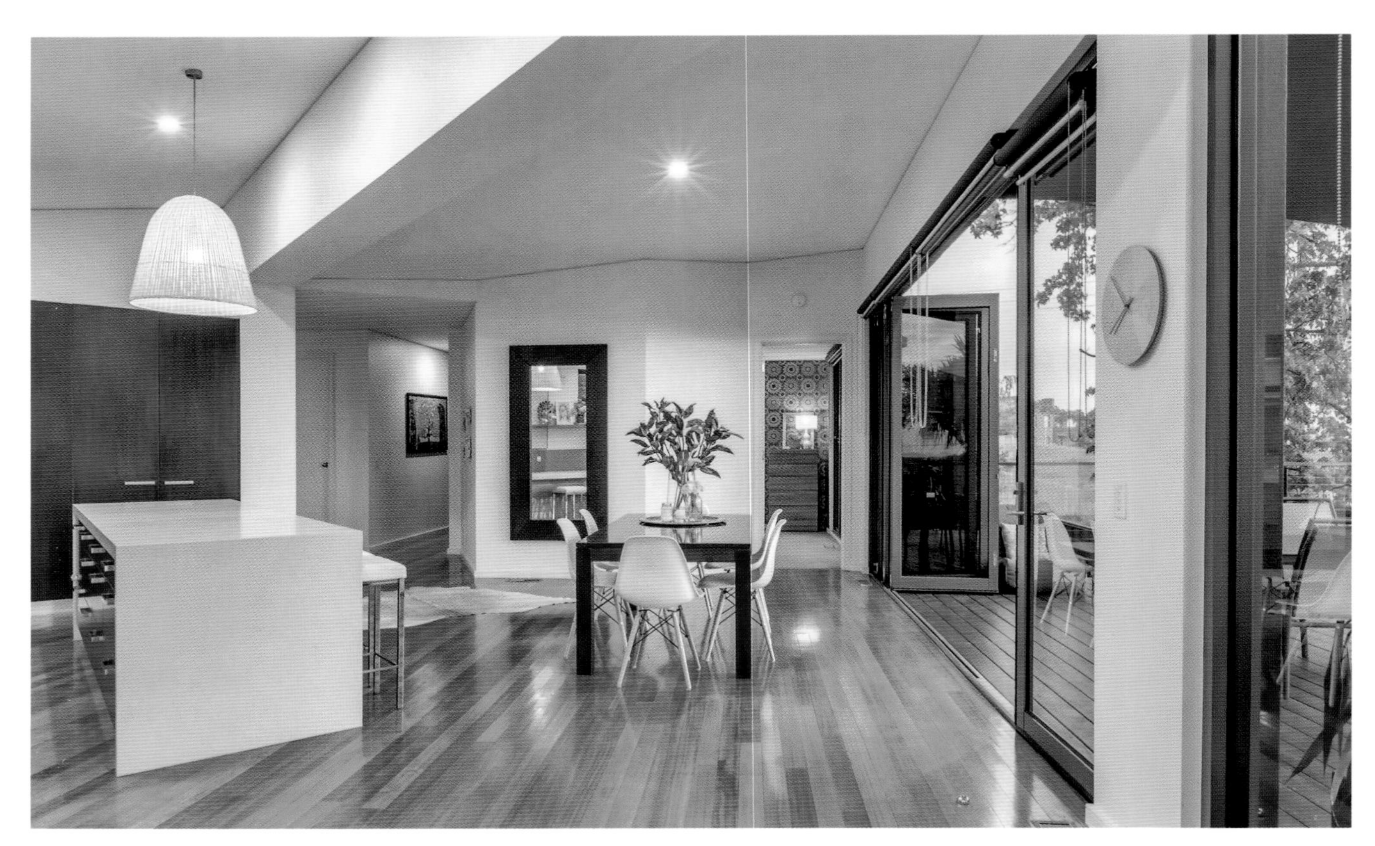

etcetera

HUE

# 切尔西三层复式住宅
# Chelsea Triplex

项目地点：英国，切尔西 ❄ Location: Chelsea, England

设计师：Alex Scott ❄ Designer: Alex Scott

设计公司：Alex Scott Porter Design ❄ Design Company: Alex Scott Porter Design

本案位于一大型联排别墅的顶楼，面朝后院的树木，感觉上远离地面的城市。本案设计的目的是为业主这个四口之家营造一个温馨舒适的空间。时间留下的锈蚀痕迹、循环再利用的材质与光滑、精致的装饰元素形成对比。成人起居室和儿童玩乐的空间紧挨在一起，让本案的结构更加紧密。

Nestled at the top of a large townhouse and oriented toward the trees of the backyard, this apartment feels remote from the city below. The goal of the project was to create a warm, inviting comfortable space that could grow with a family of four. Rustic, reclaimed elements contrast to sleek, refined ones. Grown-up and play spaces exist side-by-side to foster togetherness.

本案空间灵活多变，可以轻易地从聚会空间转为工作空间，在休闲和工作中自由转换。局部利用黑板漆装饰，可以让儿童充分发挥自己的想象力，同时也不失精致典雅。

The spaces easily shift from hosting play-dates and homework sessions, to adult entertaining and relaxing. Like the material palette, the furnishings are at once sturdy and kid-friendly and elegant and sophisticated.

Clone Wars
Whirlpool Galaxy

13:56:27

# 北欧工业风格
# Industrial Nordic Design

## 风格概述 Style Description

工业风格起源于废旧的工业厂房或仓库改造，这种改造往往保留了建筑的部分原有风貌，如裸露的墙砖，质朴的木质横梁，以及暴露的金属管道等工业痕迹。后来这种有着复古、颓废、冷峻、散发着硬朗的旧工业气息和带有特殊艺术范儿的格调就演变成了一种独特的工业装修风格。北欧工业风格是用简洁的线条搭配工业水泥地面和工业金属吊灯等工业元素，诠释自由、复古、自然的设计风格。

Nordic Industrial Style originated from transformation of old factories and warehouses. The transformation maintained the original appearance of the architectures, such as the bricks, wooden beams and metal pipes or other industrial traces. This retro, decadence, solemn style spreads the air of old factories and with an style of special art decorating. Nordic Industrial style use simple lines to coordinate with the concrete floor and ceiling lamp or other industrial style elements to present a free, retro and natural style.

## 细节设计 Detail Designing

黑白灰色系十分适合北欧工业风格，两者混搭交错又可以创造出更多层次的变化。素水泥墙质朴的色彩突出了工业风的冷酷感，打磨后散发出的光泽，特别有后现代的时尚感。与传统的家居装饰不同，工业风格装饰不刻意隐藏各种水电管线，而是透过管线的布置以及颜色的搭配，将它化为室内的视觉元素之一。除了金属家具，原木也是工业风的家具中常见的元素，许多金属材质的桌椅会用原木木板来作为桌面或者是椅面。

Black, white and gray matched well with Nordic industrial style. These colors mixed with Nordic industrial style can create a multi-level transformation. The simple color of plain cement wall highlights the cold feeling of industrial style and will be shiny after polished. It has a modern sense of fashion. Different from the traditional home decoration, industrial style decoration does not deliberately hide a variety of water pipe and electricity lines, but to make it as one of the interior elements. Except the metal furniture, wood is also a usual element in industrial furniture. Many tables and chairs with metal materials will be matched with wood.

## 设计技巧 Design Techniques

▶ 局部使用金属材质，例如厨房的操作台，灶台背景板，金属吧台等。
Use metal materials partially, such as kitchen counters, stove background plate, and metal bar, etc.

▶ 保留建筑原结构中的木梁或使用带有天然纹理和自然气息的原木做装饰。
Retain the original structure of the wooden beams and use wood with natural texture and smell.

▶ 选择带有工业风格家具单品。例如：铆钉装饰的皮质家具，金属铆钉外露的实木家具等。
Choose the furniture with industrial style, such as furniture with rivet decorative leather, wooden furniture with exposed metal rivets, etc.

▶ 选择工业风格的灯具，与休闲风格家居环境进行混搭，不突兀，却足够个性。
Select lamps with industrial style and mixed with the leisure style home environment, to manifests the special side of industrial style.

▶ 充满年代感的斑驳的老物件，很适合拿来作为工业风的元素，一个铁盒、一把剪刀、庭园工具都是很有味道的装饰。
The old objects with full of histories can be used as the industrial elements. A tin, scissors and garden tools are great tastes of the decoration.

# Itacolomi 445住宅

# Itacolomi 445 Apartment

项目地点：美国，纽约 ❄ Location: New York, America

设计师：Diego Revollo ❄ Designer: Diego Revollo

设计师 Diego Revollo 是巴西装饰界中首屈一指的设计师，他对装饰艺术和家具的挑选有自己独特的见解和手法，从天花板到墙壁都能平衡各种元素，创造出良好的视觉效果。在本案中，设计师结合纽约顶楼的特点，以 Itacolomi 445 建筑为基础，将建筑室外装饰的特色复制到室内，以灰色、黑色、中性和近黑的深蓝色为色彩基础，突出金属架构、房梁和黑色框架，以此奠定室内大空间的设计基调。所有的装饰都是有层次的精心安排，以达到内外合一的装饰效果。

From the ceiling to the walls, Diego Revollo, one of the most important names among the creators of the new Brazilian decor, offers a box that encapsulates a sophisticated selection of art and furniture, with balance and great visual effect.With New York loft air, we start from the existing, so, from the original architecture of Itacolomi 445, which is the major project asset. With features and outstanding volumes in the facade with guardrails and brises in black, the idea was to replicate these characteristics to the inside, creating a unity between external and internal architecture. In this way we emphasize all metal structure, beams and black frames that stand out in the internal gray box where integration sets the tone in large spaces. Everything was orchestrated in a hierarchy where the architecture said the entire project: Grays, blacks, neutral and navy blue almost noir.

本案严格保持了单一的结构，单色的装饰色调与棕色胡桃木实木地板形成对比，局部采用金黄色点缀，形成了时髦别致的工业风格氛围。

This monochrome structure is maintained will strictly the decor palette and contrasts only with the wood always brown in cumarus and walnut, with discrete bits of gold, which contribute to the chic industrial climate.

PIERO LISSONI

# 独具一格的阁楼

# Real Parque Loft

项目地点：巴西，圣保罗 ❄ Location: São Paulo, Brazil

项目面积：105 平方米 ❄ Project Size: 105 $m^2$

设计师：Diego Revollo ❄ Architect: Diego Revollo

本案是面积为105平方米的阁楼，位于圣保罗南部一处居民区，这里的建筑大多建于二十世纪八九十年代。本案设计面临的主要挑战是如何打通空间，在现有的有限格局中营造开阔感。设计师面对的是一个传统隔间型公寓，里面有大量分区以及与之相联的封闭式房间。本案的创意是一个以“烧水泥”为涂层的结构，既富于现代感又有一种未经修饰的简洁感。“烧水泥”用于所有表面，包括室内地面、墙面以及天花板。

The Loft, with 105m$^2$, is located in the south zone from São Paulo, in a strictly residential buildings neighborhood, basically from the 80 and 90 decade. For this reason the designers started from a traditional compartmentalized apartment with exaggerated number of divisions and closed rooms if related to its area. The main challenge was to open the space, bring the sense of amplitude within the existing structural limitations. he idea of a box with just a coating, burned cement, would bring the contemporary aspect and look like "clean" without amendments or interruptions and would be applied on all surfaces such as floors, walls and ceilings.

根据天然水泥的色调，设计师在选择其他装饰材料的时候做了特别的考虑。在浴室中，设计师选用了天然的绿洲蓝石灰岩。该材料的色调与水泥的相似，可替代水泥用于浴室地板和雕刻而成的洗手盆。

TA particular care has been taken into account in choosing old and close by the natural cement's tone. In the baths, the designers used the natural Oasis Blue limestone for the slabs. The limestone is with a similar tone of cement and employed only as an alternative to cement to be more appropriate for slabs and carved sinks.

家具与室内设计中的布艺选用的依旧是天然材料，例如天然亚麻与仿古皮革，突出温暖和舒适的感觉，淡化水泥带来的冰冷感。最终，该项目成了一个简约宽敞并且极其宜居的阁楼公寓。

The furniture and interior design continues with the choice of textiles as the natural linen or the distressed leather and prioritizes the warm touch and comfort always against the coldness of the cement box. The end result is a loft without excesses, spacious and extremely pleasant to live.

冰冷或过于现代的效果都不会令业主满意，他寻求的是一种优雅的氛围，同时兼具舒适与热烈的感觉。为此，设计师建议采用水泥和栗棕色的原木相结合，这样既可“点燃”环境氛围又能提升装饰效果。入口、餐椅、阳台椅等处采用的是高强度巴西硬木龙凤檀，从而更加凸显厚重感。在因承重或移动而不宜使用硬木的情况下，设计师选用了极具个性的红檀，该木材拥有与黄檀类似的设计感，而黄檀则是二十世纪五六十年代巴西家具的首选木材之一。

For the owner, a cold or too modern result wouldn't please him, he searched an elegant atmosphere but also comfortable and "hot". The suggestion of the office was the alliance of the cement and the natural wood in a reddish chestnut brown tone to "heat up" the environment and that would add value in decorative point of view. In some places such as the entrance, dining bench and the balcony seat, the Cumaru wood, a Brazilian's hardwood with high resistance was used by rulers to make the wood "weigh" even more. Where the use of solid wood wasn't viable either by weigh or by the natural movement, the designers chose for the Pau Ferro sheet, a wood with enough personality and a similar design to the Jacarandá, one of the main wood used in furniture production peak in Brazil in the 50 and 60 decade, for example.

# 维斯林工作室

# PplusP Studio

项目地点：中国，香港 ❄ Location: Hong Kong, China

项目面积：223 平方米 ❄ Project Size: 223 $m^2$

设计师：廖奕权 ❄ Designer: Wesley Liu

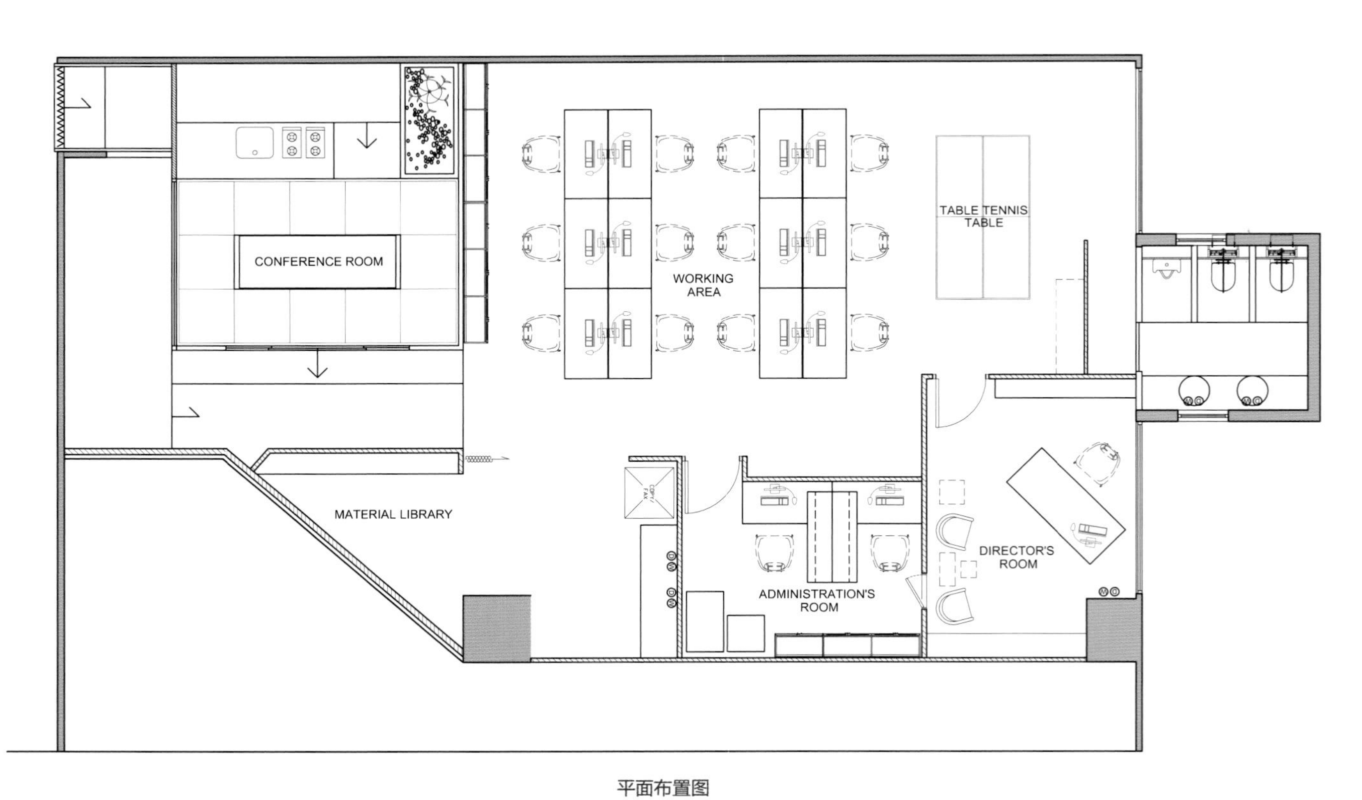


平面布置图

Apartment Plan

本案是由旧厂房改造而成的设计师工作室，占地面积 223 平方米。入门的走道被设计为迂回曲折的走廊，工作区开阔明亮，还能看见晴空下一片利落的都市天际线。设计师亲自动手，以中银大厦地标建筑为基础，描绘大家熟悉的城市面貌，其他大厦的轮廓则都是设计师的创作，包括建筑物的轮廓、每栋大厦的装饰图案及着色等，有斑驳色块、有立体穿击的痕迹，统统构成了这和谐又独树一帜的“城市天际线”主题壁。最后将开放的天花板涂成天蓝色，且在边界处做不规则的渐变色效果，让员工们恍如置身于城市中，而不是在一般的公式化的公司从事创意设计。

With an area of 222.97m$^2$, the project is a designer studio innovatively transformed from an old factory building. The hallway in the entrance is a twisted one. There is a broad and bright work area, where you can see clear city skyline under a sunny sky. The designer took the white wall as a canvas and drew the landscape of the city with skyline. With the landmark building -- the Bank of China Tower as the basis, he depicted a city familiar to all. The mottled color lumps and the traces of three-dimensional breakthrough constitute the harmonious and unique theme wall featuring city skyline. Moreover, the open ceiling was painted with sky blue, and irregular color effect characterized gradual change was created where two walls met, so the workers feel that they worked in the open air rather than in an ordinary office.

设计师在工作区与窗之间设置了一张乒乓球桌，除了发挥打球功用，还可以让员工们在此开内部会议，不必每次都去会议室，是个别具创意的点子。墙面用红砖铺贴，营造旧式风貌，搭配罗马数字大钟及复古式吊灯。

Another feature of the area is the table tennis table between the work area and the window. The large table, which seems to occupy much space, can be used for the sport and acted as a meeting table; hence, it is unnecessary to hold all meetings in the meeting room. This is a creative idea. The area is furnished with red bricks to create an ancient style, decorated with a big clock with Roman numbers and a vintage pendant.

入门转右，曲折的走廊铺设粗糙木板和榻榻米，墙角饰以水滴形吊灯，散发着自然的气息。墙上悬挂的卷轴字画来自苏州，是由国宝级书法家为工作室专门创作。

Turning right after entering the room, you can see a twisted corridor paved with rough wood blocks and tatami. The corner is decorated with a drop-shaped pendant which conveys the sense of a vast land. In particular, the scroll of calligraphy on the wall, which comes from Suzhou.

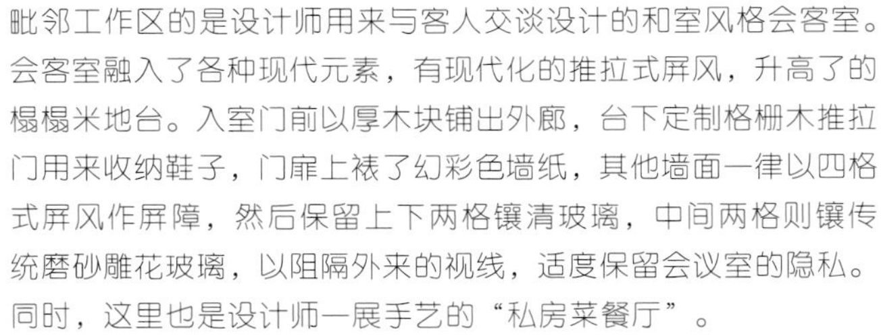

毗邻工作区的是设计师用来与客人交谈设计的和室风格会客室。会客室融入了各种现代元素，有现代化的推拉式屏风，升高了的榻榻米地台。入室门前以厚木块铺出外廊，台下定制格栅木推拉门用来收纳鞋子，门扉上裱了幻彩色墙纸，其他墙面一律以四格式屏风作屏障，然后保留上下两格镶清玻璃，中间两格则镶传统磨砂雕花玻璃，以阻隔外来的视线，适度保留会议室的隐私。同时，这里也是设计师一展手艺的“私房菜餐厅”。

The reception room is a traditional Japanese room composed of various modern elements. The modern retractable screen has hoisted the tatami hathpace. The area in front of the entrance, along with the side corridor, is furnished with a thick wood block. Under the hathpace are sliding doors for shoes storing. The door of the room is decorated with colorful wallpaper, and the room is separated from the outside by latticed screens, with the upper and lower ones made of transparent glass and the middle ones made of the traditional opaque glass with carved patterns which aim to block view from outside and maintain the privacy of the meeting room. This is also the "Private Home Cuisine Canteen" where the designer serves guests with delicious food.

大大小小的木砧板，是设计师的收藏，小盆栽也是由设计师亲手栽种的，将微缩园林植入室内。铁器、旧式茶壶、茶台，属现代工艺品，还有放在木盒上的国际象棋，各有特色。连陪衬的碗筷、酒具也都具有设计感。

The wood chopping blocks, big and small, are the collections of the designer. The potted plants were also planted by the designer who introduced a miniature garden into the room. The iron tea pot and board with ancient style are modern crafts. The modern chess on the wooden box are also distinctive. The natural Japanese bowls, chopsticks and drinking vessels are taken as foils.

设计师在公司一角辟出自己的工作室，有一面临海的大窗，工作台椅随意摆放，一对牛仔布套的沙发是设计师的收藏，沙发之间放着一个充满历史感的箱柜。办公室内设有洗手盆，是用旧缝纫机的结构做成托架，怀旧味道十足。其中一面墙采用隔音的水松板装饰，深沉的色调平衡着偌大的海景。墙上的一扇木板窗是与秘书面对面沟通的窗口。

The designer created his own studio in a corner of the company. The window of the studio faces an ocean, and the desk and chair are placed casually. The two sofas furnished with cowboy cloth are the collections of the designer. One wall of the room is decorated with soundproof Chinese cypress, and the dark color supplements the vast view of the ocean. Between the armchairs is a cabinet conveying a strong historical sense. On the wall is a wooden window for the face-to-face communication between the boss and the secretary.

White

由乒乓球桌区沿边走，不难发现“TOILET”这个招牌字样，转个弯还有“LADIES”和“GENTS” 字样，这都是设计师在澳洲搜罗的古旧精品。更醒目的是一块镶在红砖墙显眼位置的来自旧车站的告示板，也是设计师的收藏。洗手间的另一个焦点是由陶瓮缸改造的一对洗手盆，衬以圆形光环挂镜，两镜之间有个通花砖装饰的小窗。木质梳理台一旁摆上植物、烛台与饰品，让洗手间看起来像一个微缩的花园。

Walking along the table tennis table, you would see the sign of "TOILET". And taking a turn would lead you to "LADIES" and "GENTS". All the signs are the ancient artifacts collected in Australia by designer. Moreover, the billboard of an old station, also a collection of the designer, finds its perfect attractive place on the red brick wall. Another attraction in the toilet is the pair of lavabos. Between the two mirrors is a sealed window featuring hollowed latticed pattern with decorative characteristics. On the wooden table are plants, candle sticks and ornaments, making the make-up room a miniature garden.

# IG

## IG

项目地点：白俄罗斯，明斯克 ❄ Location: Minsk, Belarus

项目面积：69 平方米 ❄ Project Size: 69 $m^2$

设计师：Alexander Malinin, Anastasia Sheveleva ❄ Designer: Alexander Malinin, Anastasia Sheveleva

设计公司：INT2 Architecture ❄ Design Studio: INT2 Architecture

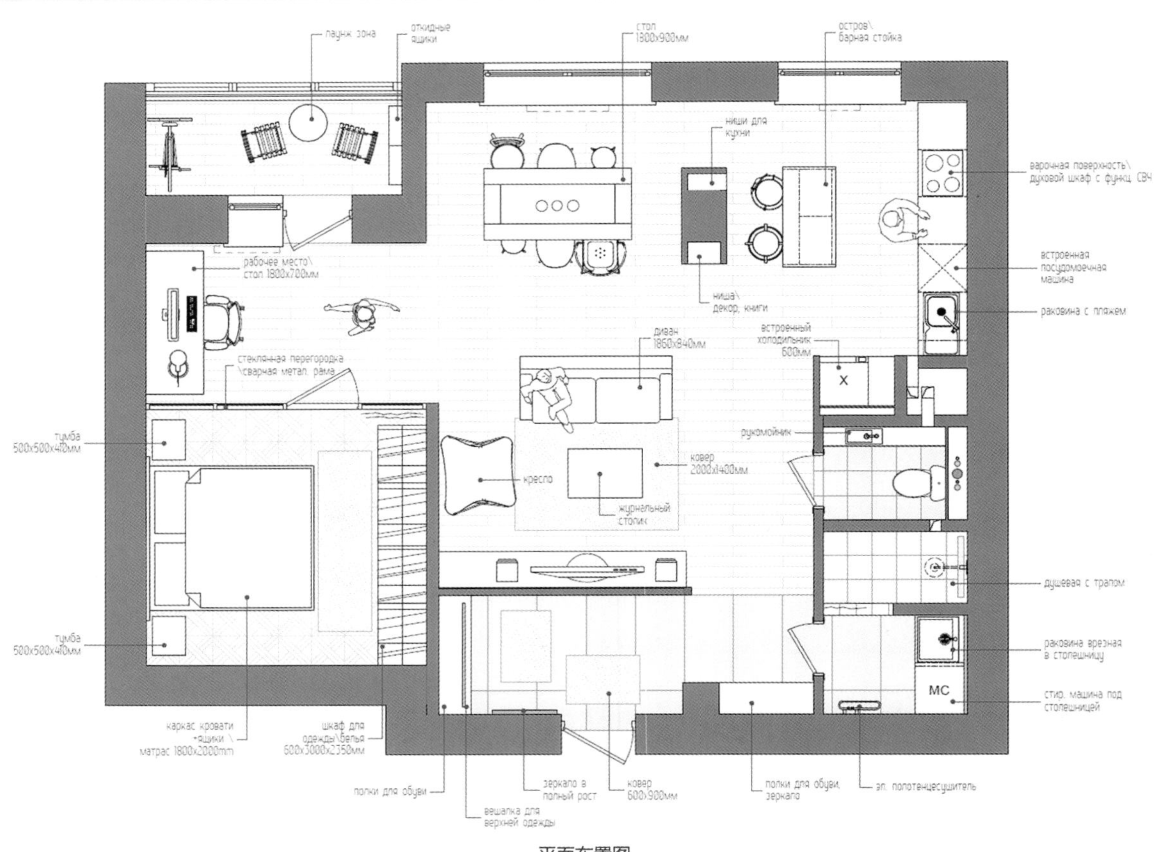


平面布置图

Apartment Plan

本案设计的主要目的是打造一个休闲咖啡吧式的家居环境，在未来几年内这将是一位年轻的 IT 专家的一个休憩场所。设计师在本案中采用了开放式的布局设计，内有独立的卧室和浴室。卧室安装了一块玻璃隔板，这样就可以把阳光引入室内。和公寓内的其他家具（如电视柜、酒吧架、餐桌、咖啡桌）一样，该隔板也是用金属焊制。此外，该项目中还有其他工业化元素：裸露在外的通风管、用复古铸铁滑轮组做成的台灯、门一样的床头板、卧室墙上镶的旧木板以及客厅与走廊之间用两种不同图案的砖做成的隔墙。

The main objective of the project was to create an interior with an anticafe atmosphere, a transit point for the young IT specialist for the next few years.The apartment is designed as an open-concept floor plan with isolated bedroom and bathroom. The bedroom has a glass partition that provides solar light access to the space. This partition will be welded using metal, as well as many other articles of furniture in the apartment: TV stand, bar shelf, dining and coffee tables. And these are not the only industrial elements in this project: open air ducts, the lamp made from the antique cast iron pulley wheel block, door as a headboard, old wood planks on the bedroom wall, the partition between the living room and the hallway made from two types of brick.

МЕРА ВЕСА
1 ч. ложка = 5 мл
1 ст. ложка = 15 мл
1 чашка = 240 мл
1 пинта = 475 мл
1 кварта = 950 мл

PiCTiONARY
Команда 1
Команда 2
SCRABBLE
MONOPOLY

ND-002

ND-002
ND-003

13:56:27

I WANT TO
BELIEVE

WAKE THE
FUCK UP

logo
IT'S NOT
ONLY
ROCK'N'
ROLL
BABY!

# HB6B

## HB6B

| | |
|---|---|
| 项目地点：瑞典，斯德哥尔摩 | Location: Stockholm, Sweden |
| 项目面积：36 平方米 | Project Size: 36 $m^2$ |
| 设计师：Karin Matz | Designer: Karin Matz |
| 摄影：Karin Matz | Photographer: Karin Matz |
| 施工：NCE Bygg | Construction: NCE Bygg |
| 木工：Retsloff Snickeri | Carpenter: Retsloff Snickeri |

本案位于瑞典斯德哥尔摩的 Heleneborgsgatan 路，2012 年被出售。在此前的 30 年，它一直被用来存放家具。以前的业主在 80 年代开始进行整修，但后来病倒了。直到他去世，该公寓一直原封不动。时间静止了，墙纸有一半已经剥落，只有几片瓷砖和一个厨房的水龙头从墙壁中冒出来，屋里没有电，有的只是一个老鼠窝似的浴室。在斯德哥尔摩这样一个房源紧缺和房价每分每秒都在上涨的城市，这个故事多少让人无法理解，也无法让人抗拒。经过设计之后，该公寓成了一件夺目之作。这是个尝试，让空间中原先的层次和故事保留下来又同时满足新故事展开的需要。

When the apartment on Heleneborgsgatan in Stockholm, Sweden was for sale in 2012, it had been used as furniture storage for 30 years. The previous owner had begun a renovation in the 1980s but fell ill and the apartment was left untouched until his death. Time had been frozen; wallpaper was half removed, only a few tiles and a kitchen faucet were sticking out of a wall, there was no electricity and a bathroom only with signs of rats as inhabitants. In a city like Stockholm with an enormous housing shortage and with every square meter increasing in price by the minute, this story was somehow impossible to understand and resist. The finished apartment is a result of a fascination for this: a try to let the previous layers and stories of a space live on and at the same time fill the requirements for the new story that will take place.

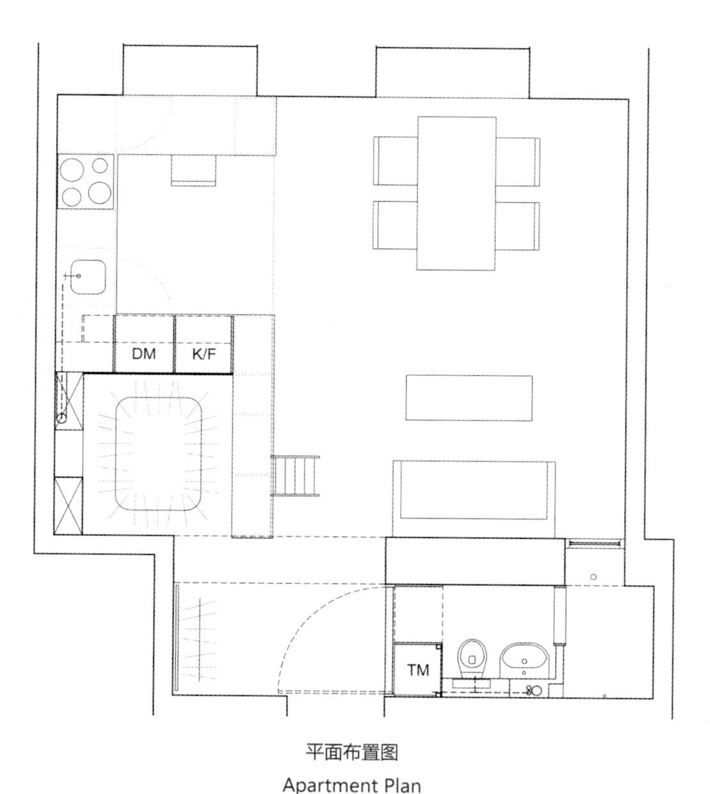


平面布置图

Apartment Plan

本案面积只有36平方米。设计师的目标就是将入住者所期盼的一切囊括其中。本案中，我们可以看到宽敞的空间、通风的环境、步入式衣帽间、日常生活所需的家用电器、大而奢华的淋浴间和浴缸、各种移动的可能性，以及一个可以任意划分的空间。

The apartment is 36 $m^2$ and the goal was to fit everything desired by the occupant. In this case: generous spaces, airy sensation, walk in closet, all appliances for everyday life, a large luxury shower and bath, different possibilities of movement, a space which could be divided whenever the occupant wanted.

设计的成果就是一个一分为二的公寓。在其中一个空间，里边的所有东西都是结构的一部分，该结构以宜家的厨房设备为基础，墙体内电路重置，表面均刷上白漆以引入和反射光线。在这里，所有的功能都压缩在一起，或居中，或上下，或里外。卧室、厨房、衣橱和储藏间融为一体。

The result is an apartment divided in two parts. One where everything is part of one structure, which is based on the Ikea kitchen units. Everything in this part is completely redone with electricity inside the walls and with all surfaces painted white in order to bring in and reflect light. Here all the functions are squeezed in on top of, in-between, under and inside each other. Bedroom, kitchen, wardrobe and storage are all one.

在另一空间中，物件都是单独放置，所有东西的外观或多或少都还保留着 20 年前的模样。原先的墙洞被补上了，松垮的墙纸和墙漆也被清理掉，新增的电缆和电源插座则裸露在墙面。

The second part is left with things free-standing with all surfaces more or less as they have been for the last 20 years. The holes in the walls have been filled in, loose wallpaper and paint taken down and electrical cables and outlets have been added running on the outside of the walls.

# 春纪的住宅

# Haruki's Apartment

项目地点：乌克兰，马里乌波尔

项目面积：35 平方米

设计师：Larisa Gusakova

设计公司： The Goort

3D 视效设计：Aleksandr Gusakov、Larisa Gusakova（The Goort）

Location: Mariupol, Ukraine

Project Size: 35 $m^2$

Designer: Larisa Gusakova

Design Studio: The Goort

3D Visualizations: Aleksandr Gusakov, Larisa Gusakova (The Goort)

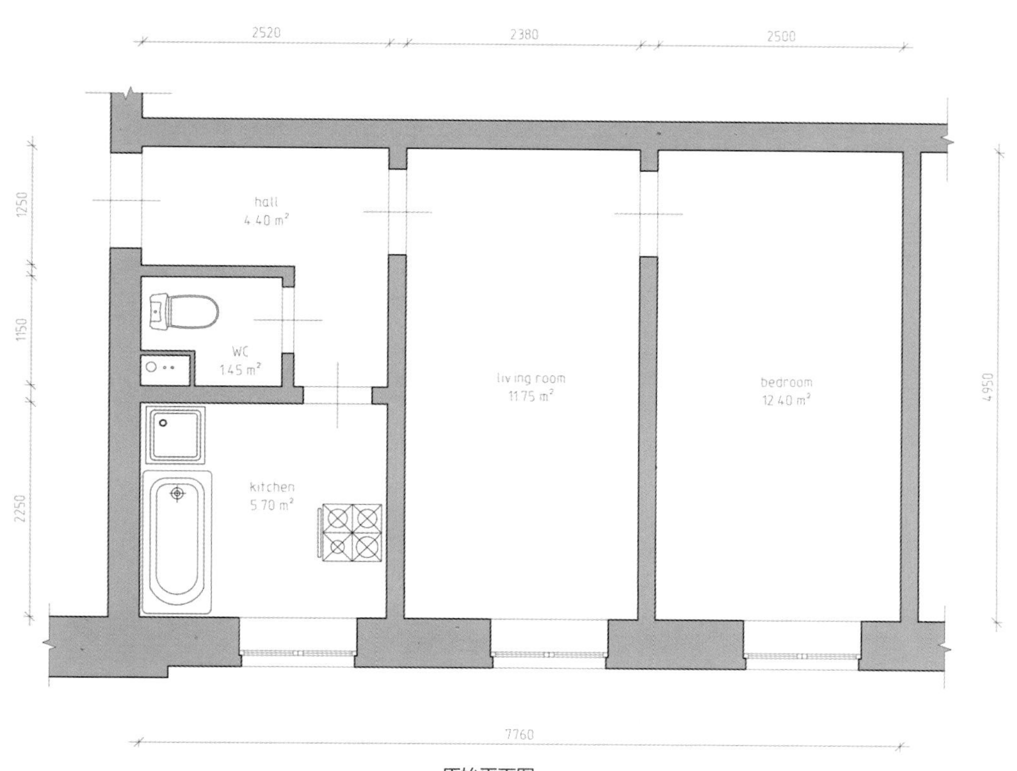


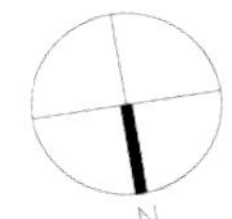


原始平面图
Original Planimetric Map

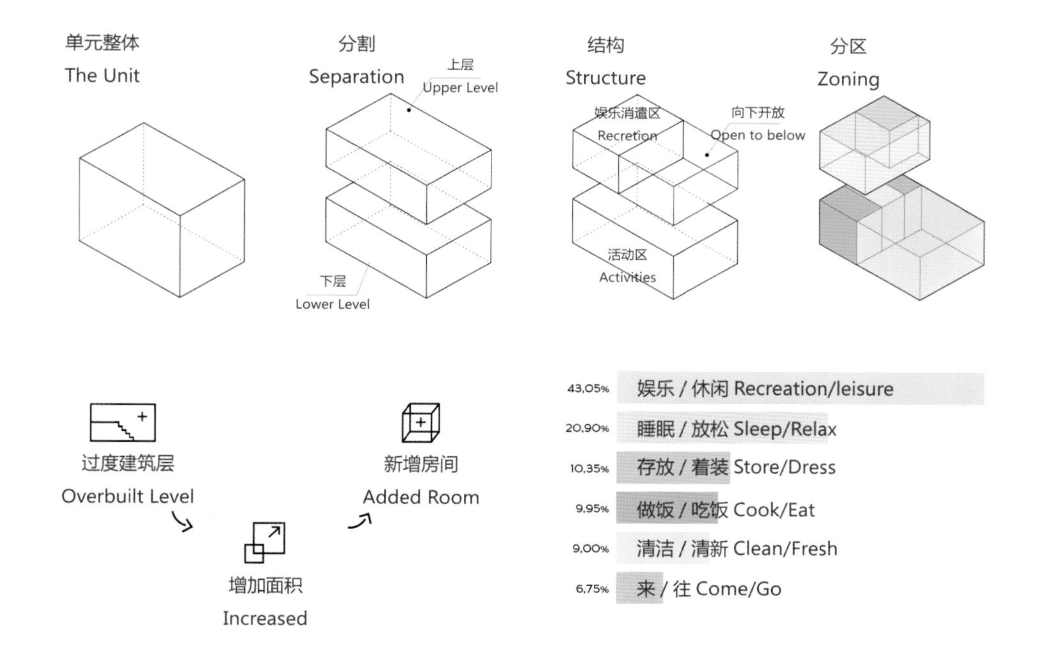


本案为单房公寓，总面积 35 平方米，位于马里乌波尔市历史中心一栋两层砖砌建筑的第一层。该市的第一家报社的印刷办公室就设在本案所在的这座公寓中。现在的业主是一对年轻的夫妇，他们选择了现代都市公寓宽敞明亮的设计，以尽可能少的家具实现尽可能多的功能。

One bedroom apartment with total area of 35m² is located on the first floor of a two storey brick building in the historic center of the city. A printing office of the city's first newspaper was situated in this apartment. The current owners, a young couple, chose the format of a modern city apartment, spacious and bright, with a minimum of furniture and a maximum range of functions.

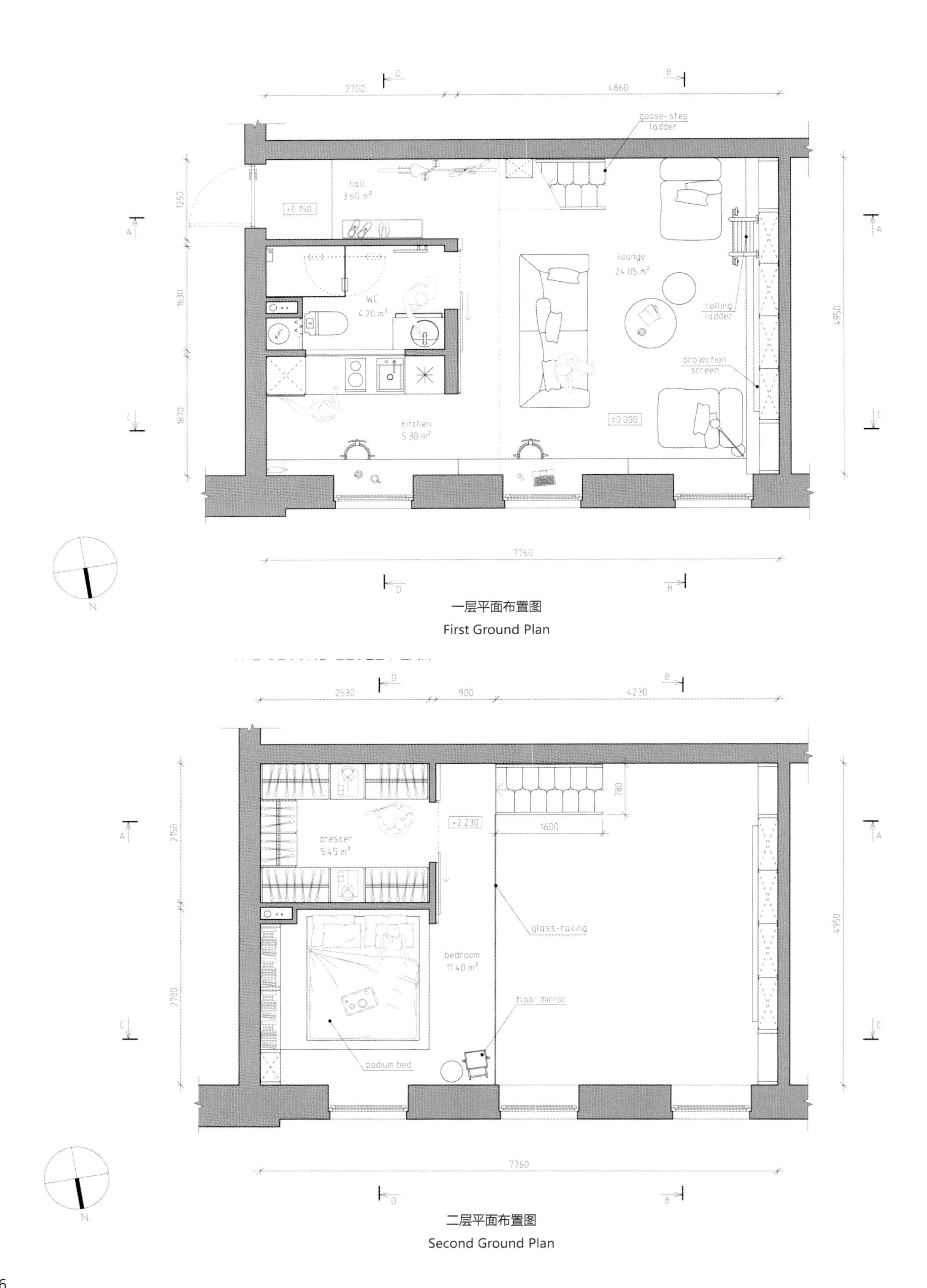


一层平面布置图

First Ground Plan

二层平面布置图

Second Ground Plan

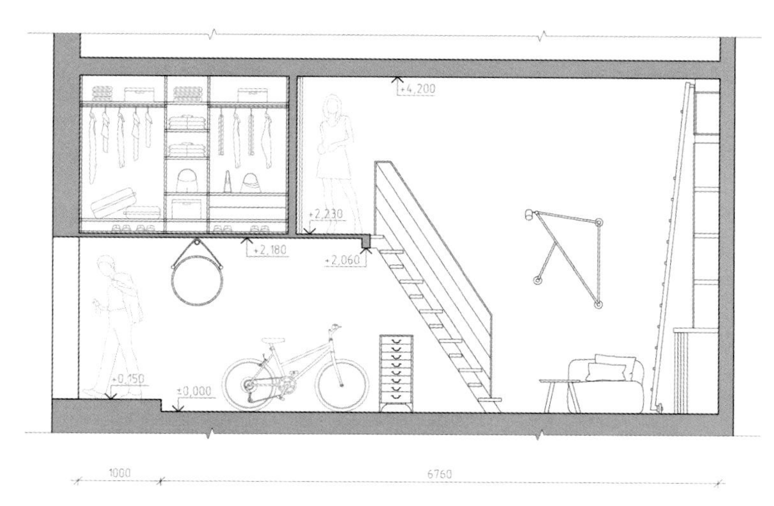


A-A 区 剖面图

Section A-A

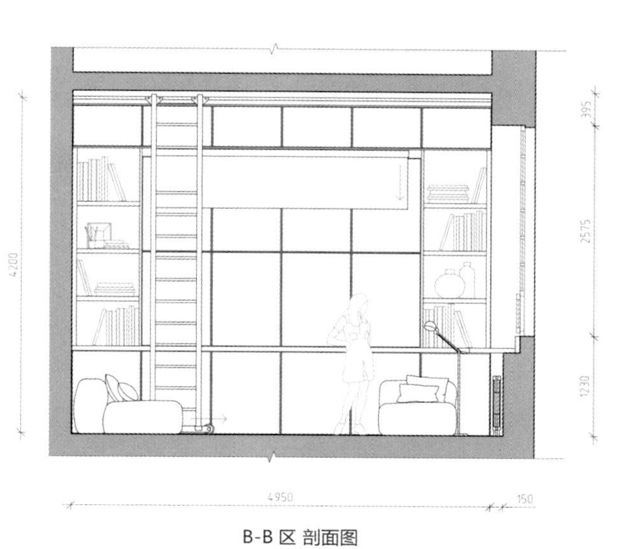


B-B 区 剖面图

Section B-B

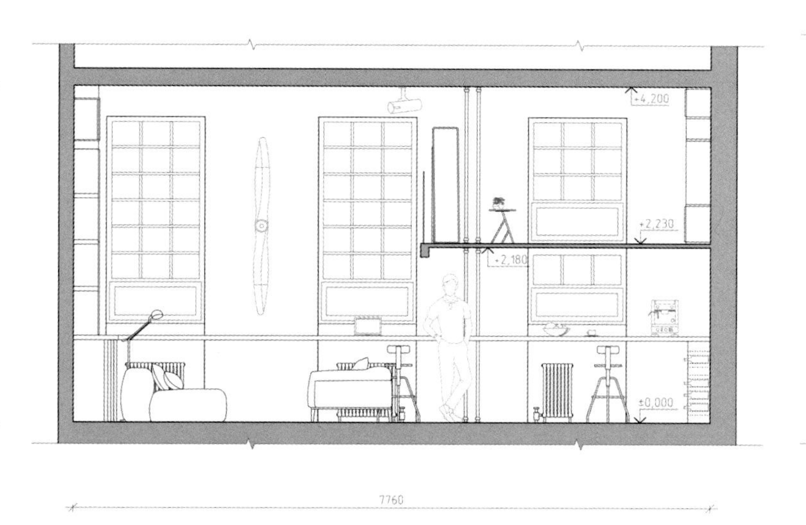


C-C 区 剖面图

Section C-C

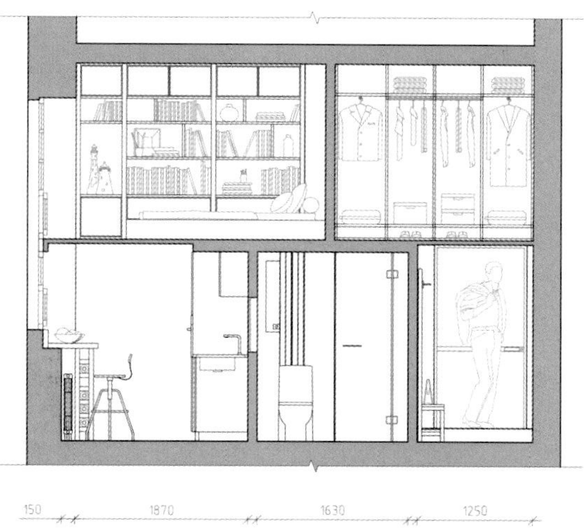


D-D 区 剖面图

Section D-D

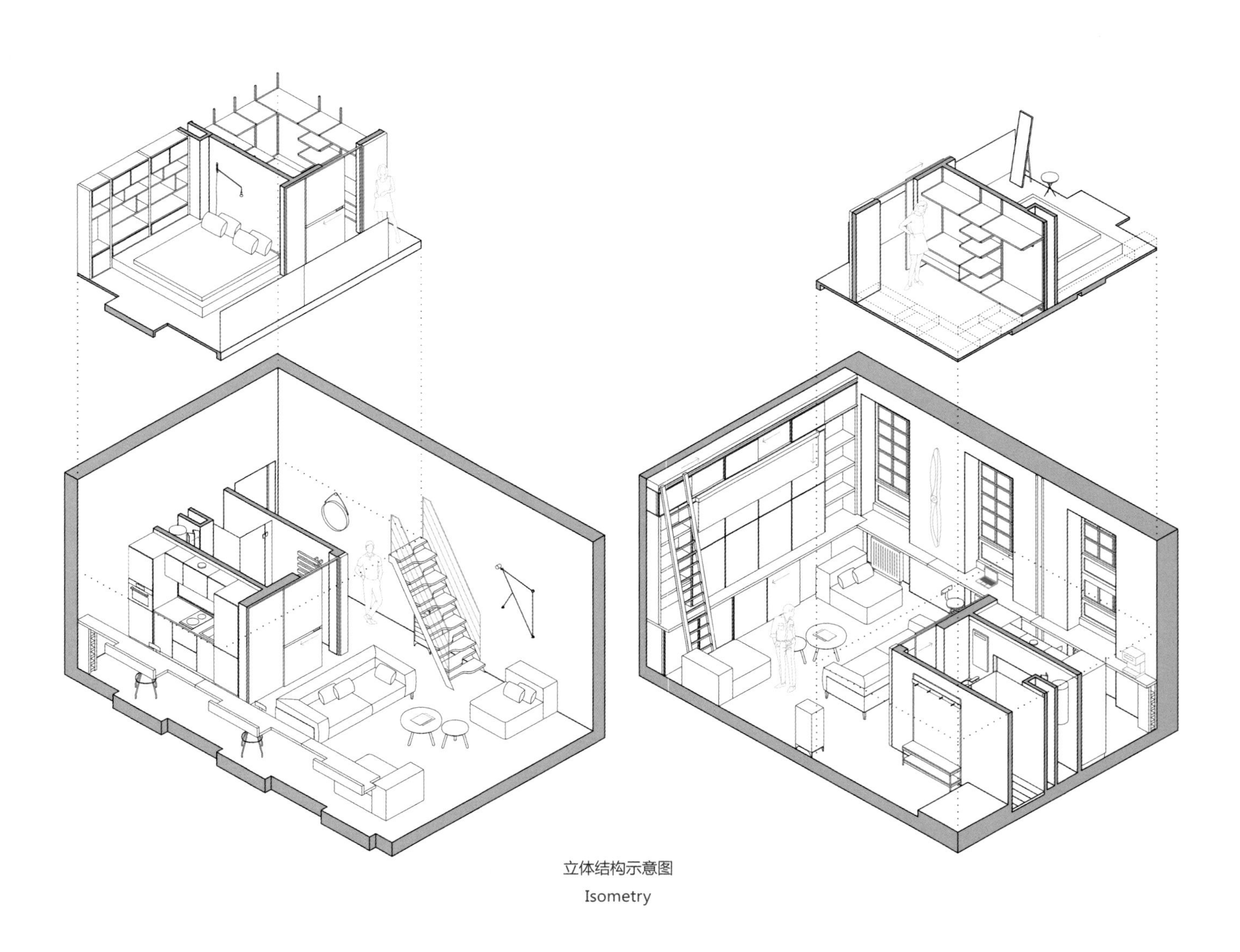

立体结构示意图
Isometry

通过拆除原先的地板（厚为 0.15 米 ~ 0.20 米的弧形木地板，分布于各个角落），将层高提升至 4 米，这是本案最关键的设计。因此，设计的重点在于垂直空间的利用。各功能也是按照房子的用途进行分布：第一层为公用空间，第二层为私密性空间，这两层通过楼梯和共用的第二盏灯联系在一起。

The main trump card, which solved all the space—ceiling height (4.0 m), which was slightly increased by the dismantling of the old flooring (curve wood floor, drop in different corners amounted to 0.15-0.20 m). Thus, main volume was resolved on a vertical rather than horizontal; functions were distributed according to the purpose of premises: the first level—representative, the second — private, the connected with the help of stairs and a common second light.

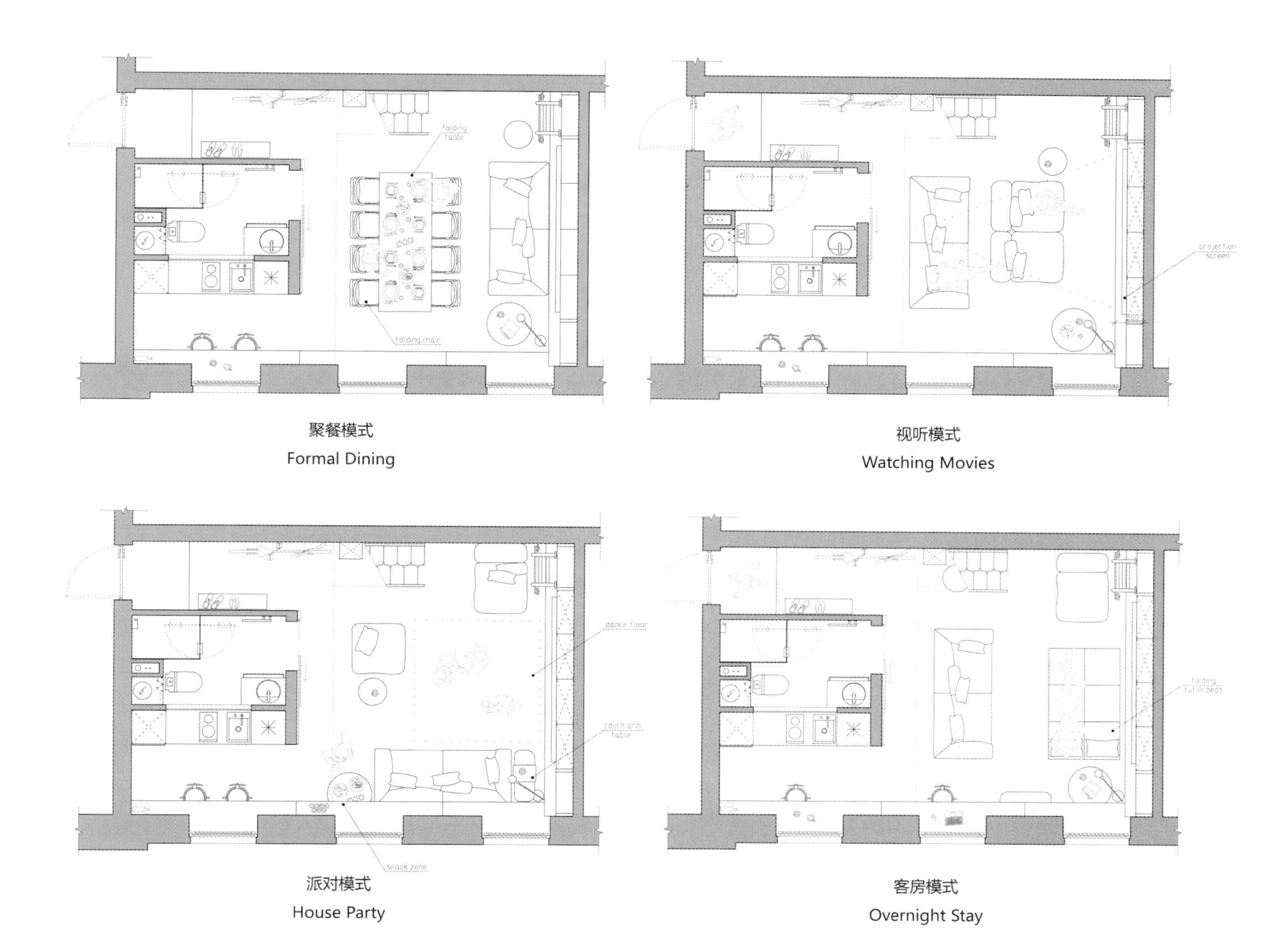


聚餐模式
Formal Dining

视听模式
Watching Movies

派对模式
House Party

客房模式
Overnight Stay

业主很喜欢社交，会经常在家里约见朋友或举行派对。整个空间最大的部分就是客厅，客厅需要具有正式用餐、自助餐、家庭派对、家庭影院以及客人临时留宿的功能，所以客厅经过设计后功能应多样化且可以轻松转换。

The owners of the apartment are very sociable people, they have a lot of friends and a lot of reasons to meet them, including meetings at home.Thus, The biggest part of the area is in a lounge zone, which actually changes its appearance very quickly and easily, depending on which task it has.

CLOSE
EYES
TO
EXIT

NO.1
NO.2
NO.3
NO.4
NO.5
NO.6
NO.7
NO.8

客厅功能改动所需要的物件都藏在一个巨大的储物柜里，这个储物柜从地板直到天花，布满了客厅的一整面墙。柜子的下半部分存放可拆卸的桌子和折叠椅，中间部分放置抱枕和床垫，其他部分用于存放私人物件，借助滚轴梯可以拿取顶层柜格的任何物品。

The required items for these modifications are located (hidden) in the cells of a huge cabinet, which occupies the entire space from floor to ceiling in the main room. Its lower part is used for hiding dismountable table and folding chairs, central—cushions and futons to sleep, everything else—for personal items, ladder roller allows to reach the uppermost cells in any part of the cabinet.

MICHELLE
MONAGHAN

MICHELLE
MONAGHAN

房间里没有单独的工作区，而这一功能则通过一条长长的窗台来实现。在厨房区，该窗台可以当作早餐和简餐的餐桌。

There is no separate working area as such, this function is carried out by a long windowsill. It is used as a table for breakfast and light dinner in the kitchen area.

DO
ALL
THINGS
WITH
LOVE
NEW YORK

EAT
WELL
TRAVEL
OFTEN

# 别墅阁楼

# Loft Vila Leopoldina

项目地点：巴西，圣保罗 ❄ Location: São Paulo, Brazil

项目面积：70 平方米 ❄ Project Size: 70 $m^2$

设计师：Diego Revollo ❄ Architect: Diego Revollo

本案为 70 平方米的阁楼公寓，位于充满现代建筑风格的巴西圣保罗维拉利奥波尔迪纳区。业主是一位年轻的音乐剧与广告演员，其现代化的生活方式体现在阁楼的设计理念中。在这里社交区域与私密区域仅一帘之隔，令空间仿佛戏剧舞台一般。整个公寓的结构，包括室内地面、墙体以及天花板，全部由“烧水泥”这种单一材质铺设。窗帘、客厅地毯、橱柜则选择了紫红色、紫色以及淡紫色，为中性的结构增添了非同寻常的色彩搭配。与“小空间搭配小家具”的理念恰恰相反，客厅放置的是一张占据两面墙的特大号沙发。这张沙发让整个客厅看起来更加开阔，效果出乎预料得好。

Located in Vila Leopoldina district on a contemporary way of living building, this 70 $m^2$, loft belongs to a young musical and advertisement actor. His modern lifestyle reflects the concept of an integrated loft, whose social and intimate areas are separated only by a curtain, which also refers to a theatrical scene. For the whole structure of the apartment, it was chosen a single material, "burnt cement", which covers floor, walls and ceiling. It was added to this neutral structure an unusual mix of shades of burgundy, lilac and purple used on the curtains, on the carpet of the living room and also in the kitchen. Contrary to the idea that small spaces ask for small furniture, it was chosen an ultra sized sofa, which takes up two walls of the living room. Unlike what people imagine, this sofa has brought immense breadth and lightness to the room.

厨房的结构以灰色为主，淡紫色的橱柜搭配黄色柜顶。黄色是餐区和备餐区的主导色，打破了原有的色度组合，使公寓显得更加时尚欢乐。此外，炉具与餐桌采用同一高度，这也不失为一个大胆的设计。

The kitchen with gray structure and burgundy cabinetry received a yellow top, which occupies the eating area and also the food preparation area. The choice of yellow broke the used shades and made the apartment even more modern and cheerful. In addition, the top on the same height for the stove and for the dining table was a bold architectural solution.

此外，私密区内还有一个隔音的小型办公室，便于屋主直接在家办公、录制音频。

Also in this area, it was created a small office with acoustic isolation that allows the owner to work and make audio recordings directly from his home.

本案设计师充分利用空间，打造出了一个实用至极又兼具美感的公寓。

The result is a well used apartment, that brings everything to a practical and actual lifestyle, without losing the beauty.

THEFASHIONBOOK
CENTURY

应本案单身业主的要求，设计师对私密区的设计旨在提升舒适感。衣柜、浴室、浴缸划分明显，地板均采用天然原木。该区域的整合营造出一种水疗馆的氛围，又或给人一种置身“浴场”的感受，对于面积仅 70 平方米的公寓来说实在是意外的惊喜。

As a request of the client who lives alone, the intimate area was designed to bring a lot of comfort. The closet, the bathroom and tub were sectored by creating a gap and also with the choice of natural wood for the floor. The integration of this area created an atmosphere of SPA or a real "bath room", unexpected for an apartment of 70 $m^2$.

# Halle A

## Halle A

项目地点：德国，慕尼黑 ❄ Location: Munich, Germany

项目面积：650 平方米 ❄ Project Size: 650 $m^2$

设计师：Sasa Stanojčić, Christina Koepf ❄ Designers: Sasa Stanojčić, Christina Koepf

设计公司：Designliga ❄ Design Studio: Designliga

摄影：Büro für Visuelle Kommunikation Innenarchitect ❄ Photographer:Büro für Visuelle Kommunikation Innenarchitect

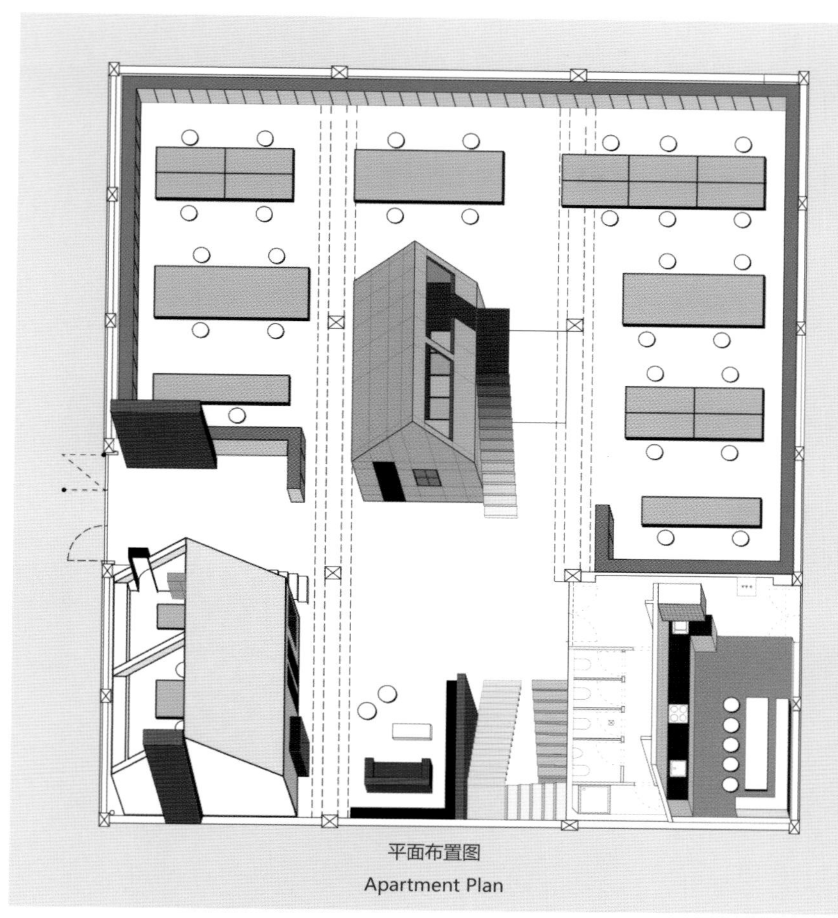

平面布置图

Apartment Plan

本案原是一座建于 20 世纪中叶的废弃老厂房，拥有 650 平方米的大厅空间，昔日机械车间的痕迹仍清晰可见。对设计师来说，这里是德国工业化的遗产，它就像一面镜子，映射工厂作业的精髓，激发人们审视工业在当代的意义。如今 Form&Code 将其改造成了自己的新办公室，由 Designliga 设计事务所设计。

A former industrial complex from the mid-20th century, with 650 $m^2$ of hall space and visible traces of its previous use as a machine shop. For Designliga, this legacy of German industrialization is a mirror reflecting the quintessence of work, and an inspiration to examine the meaning of work in our contemporary age. Designliga, took a former industrial machine shop and created a new working environment for its own staff and for Form & Code, its strategic partner in Web and application development.

大厅有一个矩形的活动空间，天花板最高逾 10 米。东、西两侧使用玻璃砖建造，整排窗户与视线平行。光线倾泻而入，给大厅带来持续柔和的光照。地板包含了斜纹样式的落叶松镶木地板和混凝土，为整个大厅加入了可视化的结构。裸露的起重机轨道和加热元件延伸贯穿整个大厅。部分室内空间采用复式楼结构，领班的办公室俯瞰大厅与入口，侧面均为砖砌墙体。

The hall has a rectangular footprint and ceilings over 10 metres in height at their highest point. The east and west sides are built of glass bricks and incorporate a row of windows at eye level. The hall is thus flooded with a consistent level of glare-free light. Flooring combines areas of cross-grained larchwood parquet with concrete, adding visual structure to the hall. Exposed crane tracks and heating elements extend throughout the entire hall. Part of the interior is two-storey; what was once the foreman's office overlooks the hall and the entrance, which is flanked by two brickwork walls.

各工作区分布在正中央黄色“房子”周围，由一块 85 米长的侧板作为界限。侧板一端用作前台接待，另一端把流通区域与工作区隔开。两个房子之间的区域构成一个“村庄广场”。从入口到会议室、从工作区到厨房，这两条轴线在此交汇。“广场”延伸至一个开放式阅读区。

Workspaces are distributed around the outside walls of the central brass-finished 'house'. They are bordered by an 85m length of sideboard, which serves as a reception desk at one end and separates work areas from the circulation space at the other. The area between the two houses forms a 'village square'. This point is the intersection of both axes of the interior: from entrance to conference rooms, and from workspaces to kitchen. The 'square' extends into an open-plan library area.

大厅保留了旧机械车间的元素，例如大型工业时钟。这些元素增添了强烈的工业时代气息，让人回想起当时所崇尚的理性、效率以及进取精神，而这些核心价值观对于营造氛围也至关重要。

Elements saved from the hall's days as a machine shop, such as the large industrial clock, add deliberate references to the industrial age and its core values of rationality, efficiency and entrepreneurial spirit that play such a vital role in creating the atmosphere.

"村庄"作为大厅设计的核心元素，给人一种归属感，也象征着易于管理的范围、亲近和个人参与。起脊房、带公园长椅的村庄广场、黄铜房周围"花园"里的开放式工作区等，共同营造出踏实宁静的氛围。厨房在建立情感纽带方面也同样重要，人们可以在这里独自或与大家共度休闲时光。厨房的特色在于其家居元素：大餐桌、单独成组的椅子、吊灯、线条匀称迷人的冰箱，以及一扇可以远眺绿地的窗户，这一切都是为了让"村民们"团结一致，共享熟悉而亲密的社区。

The 'village' is the core element in the design of the hall, generating identity and serving as a symbol of manageable scale, proximity and personal involvement. The ridge-roofed houses, the village square with park bench and the open-plan working areas in the 'garden' around the brass-walled house create a grounded, tranquil atmosphere. The kitchen is equally important in establishing a feeling of connection and is used for personal free time, together and individually. It features homely elements including a large dining table, individually grouped chairs, lamps and refrigerator with rounded, appealing lines, and a window overlooking green spaces, all serve to unite the "villagers" in a familiar and intimate community.

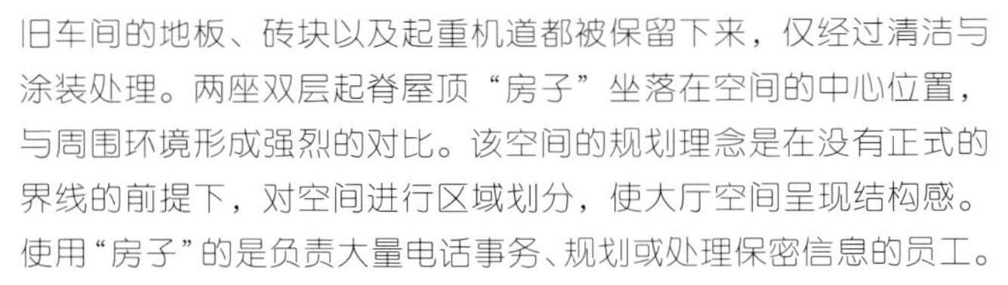

旧车间的地板、砖块以及起重机道都被保留下来，仅经过清洁与涂装处理。两座双层起脊屋顶“房子”坐落在空间的中心位置，与周围环境形成强烈的对比。该空间的规划理念是在没有正式的界线的前提下，对空间进行区域划分，使大厅空间呈现结构感。使用“房子”的是负责大量电话事务、规划或处理保密信息的员工。

The floors, brickwork and crane tracks from the former machine shop were retained and merely cleaned and painted. A pair of two-storey ridge-roofed 'house' enclosures now stand at the heart of this atmospheric setting, forming the most striking architectural contrast. As a key component of the space planning concept, they establish distinctive areas and give structure to the hall space around them without the need for formal boundaries. The 'houses' are used by staff involved in extensive telephoning, planning or working with confidential information.

U

U

楼上的两个领班办公室被改造成风格对比强烈的会议室：一个是带传统会议设备的功能房间；另一个是用于非正式会面的沙龙厅。会议室之间的实体墙已被拆除，代之以玻璃墙。

The two foreman's offices on the upper floor were converted into conference rooms with contrasting styles – a functional room with classic conference equipment and a salon for informal meetings. The solid wall between the two offices was removed and replaced with glass.

建筑的氛围、现代化、数字化的工作环境、当代人对亲密与物质的追求，三者之间动态交互，为本案的设计理念提供了灵感。创造出了一个任何人走进来都想要对它进行探索并融入其中的空间。

The design concept was inspired by the dynamic interaction between architectural atmosphere, the modern digitally focused world of work and the contemporary search for intimacy and material substance. The aim of this project is to creat a space which immediately creats the desire in anyone entering it , and to explore the place and become a part of it.

ALEX SCOTT PORTER A+D

**Alex Scott Porter Architecture + Design**

www.alexscottporter.com

Alex Scott Porter Architecture + Design PLLC is a design intensive studio founded in 2001 in order to create nuanced, natural spaces. We believe in relaxed, lived-in modernism. HOMES are the firm's life-blood. We have a sustained interest in how people really live, and create houses and apartments that can adapt and change as they grow. Our work is grounded in a sense of place, in sympathy and support of the lifestyle of our clients.

**BRUZKUS BATEK**

www.bruzkusbatek.com

Founded by Ester Bruzkus and Patrick Batek in Berlin in 2007, this internationally active architects office specializes in designing hotels, offices, shops, restaurants and private housing and particularly in the detailing of high-quality interiors. The office has also completed successful exhibition and furniture design projects, including developing lines of furniture for manufacture and retail. Ester Bruzkus and Patrick Batek have an integrated approach to design.

**Con3studio**

www.con3studio.it

Formed in 1993 by Elena Belforte and Giusi Rivoira, Con3studio is located in the epicentre of Turin's urban renewal: Regio Parco. Con3studio offers global services thanks to the experience acquired in many years of practice. We specialize in residential design and handle various different projects in both size and environments. Con3studio collaborates with a wide network of professionals and businesses, providing individual consultations and "turnkey" services.

**DESIGNLIGA**

www.designliga.com

Designliga is a bureau for visual communication and interior design. Founded in Munich in 2001 by product designer Saša Stanoj i and communication designer Andreas Döhring, the office has grown into a team of experienced designers, consultants, architects and interior designers. We work in a cross-disciplinary manner and communicate content through design. They operate in the fields of consumer and luxury goods, fashion and lifestyle, environment and culture.

**Diego Revollo**

www.diegorevollo.com

Diego Revollo began his studies in Civil Engineering in the Polytechnic School of the University of San Paulo in 1994. He joined the Mackenzie Presbyterian University in 1997 and graduated in Architecture and Urbanism in 2001. He began his career in 2000 working with Roberto Migotto and in 2007 opened his own office. He is currently part of the board of the magazine Casa e Comida, from Editora Globo.

**DISC Interiors**

www.discinteriors.com

Founded in 2011 by interior designers Krista Schrock and David John Dick, DISC Interiors has been hailed for creating warm modern designs that exude an "enviable California vibe." Inspired by natural materials, tonal color palettes, and the distinctive work of artisans and craftspeople with whom they work, the duo strives to create homes that are as stylish and refined as they are effortless and comfortable.

**French & French Interiors**

www.frenchandfrenchinteriors.com

We are a family-based company with a deep understanding of how to make a house feel like home by transforming its interior. We have in-depth experience in interior design gained through traveling to many countries the world over. Our unrivaled skills in carpentry and woodwork ensure every piece of furniture, we create for your personal space, will be an extension of your personality.

**Galleria del Vento**

www.galleriadelvento.com

Galleria del Vento is an Interior Design studio founded in 2009 by Carlo e Alessandro Colciago, designers and craftsmen. Based in Monza (northern Italy) the studio offers interior design solutions mixing bespoke furniture made up in our workshop with northern europe design, unique pieces of art and antiques. High quality standards are followed through the entire process in order to satisfy all customer needs in terms of comfort, functionality and durability.

**INT2 Architecture**

www.int2architecture.ru

INT2architecture - young active multi-diciplinary team of architects, designers, artists and photographers and based in Moscow and Saint-Petersburg, Russia. Founded by Alexander Malinin & Anastasia Sheveleva. The main line of work - private housing design, interior design of private and public spaces, landscape and industrial design, urbanism.

**Karin Matz Arkitekt**

www.karinmatz.se

Karin Matz is an architect from Stockholm, Sweden.

Since 2009 Own practice as Karin Matz Arkitekt

Member of the Swedish Architecture Collective Svensk Standard

Member of the Swedish Architecture association (Sveriges Arkitekter)

**PplusP Designers**

www.ppluspdesigners.com

PplusP Designers Ltd (P+P) is a young and rising design firm based in Hong Kong. Our practice produces high quality and innovative solutions through the synthesis of interior design and project management. Led by Wesley Liu, P+P brings together the creative mind of Wesley Liu and his experience in various design fields. The team includes architects, interior designers and graphic artists whom all share a common passion for design.

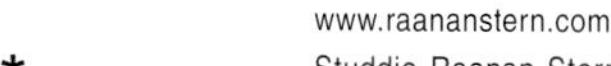

**Raanan Stern**

www.raananstern.com

Studdio Raanan Stern is a young Tel Aviv based design studio, developing architectural and interior projects in a wide range of environments ranging from small to large scale that engage in issues of creativity and know ledge, realism and imagination, function and culture. Studdio Raanan Stern gives full architectural service, from identifying spaces and potential, design and preliminary sketches, technical work plans, procurement, interior design, branding and full on - site architectural supervision. For them, each project is a new and exciting challenge.

**Spazio 14 10**

www.spazio1410.com

Spazio 14 10 is a creative laboratory of architecture, graphic design and interior design low cost, focused on the themes of recycling and ecological building materials, offering quality services with a very attractive price, thanks to the specialized knowledge and experience of the designers

It was born in 2012 from sharing ideas and passions of three graduates in Architecture and Design in Rome(Italy), in order to investigate and reinvent living spaces, from houses to cities.

**Superpozycja Architekci**

www.superpozycja.com

SUPERPOZYCJA Architectural Design Studio was founded in Katowice in 2011 by two architects and interior designers , Dominika Trzci ska and Michał Kotłowski. They specialize in building and interior design and have a detail-focused approach and a comprehensive understanding of their clients' needs. They provide their support to go through all the problems you may encounter during building or renovation works and assistance on every stage of the project's execution.

**The Goort**

www.thegoort.com

The Goort it's a creative team, which consists of an architect (Aleksandr Gusakov) and interior designer (Larisa Gusakova) from Ukraine, that specializes in contemporary laconic interiors, focusing on functionality, the relevant to place and time. The Goort likes working with a small spaces, besides the micro, apartments are very popular right now. We have more than half of the country lives in a modest living conditions and the theme of saving space, the right of his organization, new technologies relevant.

**White Elk Interiors**

www.houzz.com.au/pro/whiteelkinteriors

At White Elk Interiors we create warm, striking and functional spaces that evoke the simplicity of Nordic design. Eclectic and earth-conscious, our residential interior and hospitality fit-outs reflect our flair for reinventing contemporary spaces. For a personalized White Elk Interiors design and styling consultation please contact our Canadian studio. We work closely with our clients, transforming their visions into spaces of beauty and practical purpose.

图书在版编目（CIP）数据

室内设计风格详解．北欧 / 徐士福等主编．-- 南京：江苏凤凰科学技术出版社，2016.7
ISBN 978-7-5537-6442-9

Ⅰ．①室… Ⅱ．①徐… Ⅲ．①室内装饰设计－图集 Ⅳ．①TU238-64

中国版本图书馆 CIP 数据核字（2016）第 125575 号

## 室内设计风格详解——北欧

主　　编　徐士福　陈炬　张崟　陈加强
项目策划　凤凰空间/宋君　陈尚婷　叶广芊
责任编辑　刘屹立
特约编辑　宋君

出版发行　江苏凤凰科学技术出版社
出版社地址　南京市湖南路1号A楼，邮编：210009
出版社网址　http://www.pspress.cn
总　经　销　天津凤凰空间文化传媒有限公司
总经销网址　http://www.ifengspace.cn
印　　刷　广州市番禺艺彩印刷联合有限公司

开　　本　889 mm×1 194 mm　1/16
印　　张　16.5
版　　次　2016年7月第1版
印　　次　2019年2月第5次印刷

标准书号　ISBN 978-7-5537-6442-9
定　　价　278.00元（精）